Ergebnisse der Physiologie

Biologischen Chemie und experimentellen Pharmakologie

Reviews of Physiology

Biochemistry and Experimental Pharmacology

68

Herausgeber / Editors

R. H. Adrian, Cambridge · E. Helmreich, Würzburg
H. Holzer, Freiburg · R. Jung, Freiburg · K. Kramer, München
O. Krayer, Boston · F. Lynen, München · P. A. Miescher, Genève
H. Rasmussen, Philadelphia · A. E. Renold, Genève
U. Trendelenburg, Würzburg · K. Ullrich, Frankfurt
W. Vogt, Göttingen · A. Weber, Philadelphia

With 12 Figures

Springer-Verlag Berlin Heidelberg GmbH 1973

ISBN 978-3-540-06238-7   ISBN 978-3-540-38530-1 (eBook)
DOI 10.1007/978-3-540-38530-1

© by Springer-Verlag Berlin Heidelberg 1973
Ursprünglich erschienen bei Springer-Verlag Berlin Heidelberg New York 1973.
Softcover reprint of the hardcover 1st edition 1973
Library of Congress Catalog Card Number 62-37142.

Universitätsdruckerei H. Stürtz AG, Würzburg

Inhalt/Contents

Mitarbeiter/List of Contributors

BURGER, M. M., Prof. Dr., Biochemisches Institut, Biozentrum der Universität, Klingelbergstraße 70, CH-4056 Basel

FLEMING, W. W., Prof. Dr., Dept. of Pharmacology, West Virginia University, Medical Center, Morgantown, WV, 26506/USA

McPHILLIPS, J. J., Dr., Dept. of Pharmacology, West Virginia University, Medical Center, Morgantown, WV, 26506/USA

OWREN, P. A., Prof. Dr., Institute for Thrombosis Research, University Hospital, Oslo/Norway

STORMORKEN, H., Prof. Dr., Institute for Thrombosis Research, University Hospital, Oslo/Norway

TURNER, R. S., Prof. Dr., Biochemisches Institut, Biozentrum der Universität, Klingelbergstraße 70, CH-4056 Basel

WESTFALL, D. P., Dr., Dept. of Pharmacology, West Virginia University, Medical Center, Morgantown, WV, 26506/USA

The Mechanism of Blood Coagulation

P. A. OWREN and H. STORMORKEN*

With 1 Figure

Table of Contents

The biochemistry and physiology of blood coagulation have been subjected to intensive studies in recent years because of the enormous clinical importance of this system. Blood coagulation plays an important role in haemostasis and thrombosis by taking part in a complex interplay with blood platelets and the vessel wall. A wealth of new information has been collected which has increased our knowledge of fundamental mechanisms, improved diagnosis and treatment of haemorrhagic conditions, thrombosis and intravascular coagulation and resulted in new therapeutic agents. However, many problems remain unsolved.

* The University Institute for Thrombosis Research, Rikshospitalet, Oslo, Norway.

History

The early history of blood coagulation has been reviewed by Wöhlisch (1929), Quick (1942), Owren (1947), Jorpes (1954) and Biggs and Mac-Farlane (1962). Buchanan in 1845 first demonstrated that serum has coagulant activity and Schmidt in 1861 extracted the "fibrin ferment" (thrombin) from serum by alcohol precipitation. In 1872 Schmidt presented indirect evidence that thrombin is formed from an inactive prestage (prothrombin) in the circulating blood.

Denis in 1859 precipitated a soluble prestage of fibrin with sodium chloride and Hammarsten (1879, 1899) further developed the salt precipitation method for fibrinogen. He demonstrated that isolated fibrinogen was quickly clotted by Schmidt's "fibrin ferment". Arthus and Pages (1890) showed that calcium is necessary for coagulation.

The accelerating effect of tissue extracts has been known since de Blainville in 1834 induced intravascular coagulation by intravenous injection of brain extract and Buchanan (1845) accelerated coagulation of diluted salt plasma by fresh lymph gland extract.

Morawitz in 1904 and 1905 correlated these observations with his own experiments into the following simple two-stage concept, which is now known as the classical theory of blood coagulation: Prothrombin is converted to thrombin by thrombokinase and calcium. Thrombin converts fibrinogen into fibrin.

Morawitz introduced the term thrombokinase for the tissue factor and suggested two main sources: By tissue damage, it is released from the tissue cells. In blood which is carefully drawn so as to exclude tissue juice, it is released from the platelets by contact with a foreign surface. This idea is the first forerunner of the modern concept of the extrinsic and intrinsic pathway of coagulation.

The existence of prothrombin remained hypothetical until Mellanby in 1909 isolated a substance from plasma which was not coagulant itself, but which could be converted into thrombin.

The classical theory was not seriously challenged for 40 years. It seemed to be strongly confirmed as late as 1935–1940, by the clinical experiences with Quick's one-stage prothrombin time test which was based on the classical concept.

In 1943, the shortcomings of the old theory became obvious by the observation of Owren (1943, 1947a) of a haemorrhagic state with a very prolonged Quick's prothrombin time, in which it could be shown that prothrombin was normal, but it did not convert to thrombin by tissue thromboplastin and calcium. Study of this patient led to the discovery of the first new clotting factor, which was termed factor V (Owren, 1947a). This discovery initiated a new era of coagulation research which resulted in the discovery of several new clotting factors and to an explanation of the clotting defect in haemophilia.

During the last 10–15 years most research has centred on the biochemistry of the clotting factors, their purification and characterization and the enzyme kinetics involved in the clotting mechanism. Biological, biochemical and immunological methods for the determination of blood coagulation factors and components have been developed and introduced in clinical practice. It has been established that proteolytic reactions dominate the coagulation process, and protein and peptide chemistry is applied extensively in present day studies on blood coagulation. The theory of the coagulation mechanism has been modified repeatedly during the last 25 years in order to comply with the continuous accumulation of new evidence.

Because of the chaotic state of the literature dealing with the nomenclature of the many clotting factors, the International Committee on Blood Clotting Factors was established in 1954 with the purpose of obviating Babel and developing a common symbolic language. The system introduced by OWREN in 1947, using Roman numerals, was adopted and has been generally accepted by the scientific and medical world. In 1958, after four years of work in evaluating the evidence for the existence of the various factors and their biological and physico-chemical properties, Roman numerals were assigned to factors I through IX, with the exception of VI. The number of recognized and well characterized factors has since been increased to XIII. The number VI has unfortunately remained unassigned, because it was originally used for the prothrombin converting principle (prothrombinase) (OWREN, 1947). The clotting factors which have been assigned Roman numerals, have been thoroughly studied and characterized. Each of them is known to cause a haemorrhagic disease by absence or abnormality of the molecule, with the exception of Thromboplastin, calcium and factor XII. The Roman numerals and synonyms which have been used most commonly are given in Table 1. The Roman numerals refer to the state of the clotting factor in plasma, which most often represent an inactive precursor. The activated form of a clotting factor is indicated by the letter "a" following the Roman numeral. The existence of these twelve factors as separate and "original" clotting factors has been generally accepted, except by the school of SEEGERS (for references see: SEEGERS, 1962, 1967) who has advocated the view that factors VII, IX and X are derivatives of prothrombin and not original clotting factors. The three activities were termed autoprothrombin I, II and III and the activity corresponding to activated factor X was termed autoprothrombin C. It seems now to be evident that the purified prothrombin preparations originally prepared by SEEGERS and co-workers were contaminated with factors VII, IX and X. Due to low concentrations and due to their similar properties they had escaped detection. It has recently been shown that chromatography

Table 1. List of Synonyms

Factor I	Fibrinogen
Factor II	Prothrombin
Factor III	Thromboplastin (Tissue thromboplastin, Tissue extract)
Factor IV	Calcium
Factor V	Proaccelerin, labile factor, accelerator globulin
Factor VII	Proconvertin, Serum prothrombin conversion accelerator (SPCA), Autothrombin I, Stable factor
Factor VIII	Antihaemophilic globulin (AHG), Antihaemophilic factor A, Platelet cofactor I
Factor IX	Plasma thromboplastin component (PTC), Antihaemophilic factor B, Christmas factor, Platelet cofactor II, Autoprothrombin II
Factor X	Stuart-Prower factor
Factor XI	Plasma thromboplastin antecedent (PTA)
Factor XII	Hageman factor (HF)
Factor XIII	Fibrin stabilizing factor (FSF), Fibrinase

of these prothrombin preparations on DEAE-cellulose can remove these factors (ALEXANDER, 1958; MAMMEN, 1971). The differences in the interpretation of findings therefore, now seem to be largely reconciled.

Biochemistry and Physiology of the Blood Clotting Factors

Factor I (Fibrinogen)

Fibrinogen (factor I) is the plasma protein clotted by thrombin. Its concentration in plasma is 2–4 g per litre. It is a globulin-type of protein containing between one and five percent carbohydrate (BLOMBÄCK, 1958).

Fibrinogen is isolated from plasma by precipitation with salts (25 % saturated ammonium sulphate, 50 % saturated sodium chloride), alcohol or ether. The alcohol fractionation method of plasma proteins by various concentrations of alcohol at low temperatures and ionic strengths, was introduced by COHN et al. (1946). COHN's fraction I contains only about 50 % fibrinogen. BLOMBÄCK and BLOMBÄCK (1956) purified this fraction by washing with 1 M glycine which dissolves other proteins but not fibrinogen. They obtained samples of very high purity, claimed to have "100 % clottability". KEKWICK et al. (1955), who developed the ether precipitation method, also prepared purified preparations with 97 % of the nitrogen in fibrinogen recoverable as fibrin nitrogen. The remaining 3 % was accounted for by the fibrinopeptides split off by thrombin.

Other methods of isolation have been developed recently. A preparation by a double salt complex described by BROWN and ROTHSTEIN (1967) yields a preparation of very high purity. BERGSTRØM and WALLÉN (1961) have described a method for the preparation of fibrinogen free from plasminogen activity.

Human fibrinogen has been reported to be composed of a mixture of different forms with different solubility characteristics and in its reactivity to thrombin (MOSESSON, 1970). Methods have been developed for the isolation of these heterogenous fibrinogens, and of subfractions of fibrinogen-like proteins.

It is now recognized that the cold insoluble fraction of fibrinogen is formed by limited thrombin action (COHEN et al., 1966) and the fractions of higher solubility by incipient degradation by plasmin (MOSESSON et al., 1967). Certain of the fibrinogen species however undoubtedly are a reflection of in vivo metabolism (MOSESSON, 1970). SHERMAN et al. (1968) have demonstrated in the rabbit that the transformation of radioactively labelled low solubility fibrinogen to high solubility fibrinogen, represents a major metabolic pathway for fibrinogen.

The molecular weight, calculated on the basis of diffusion and sedimentation data is 320000–340000 for both bovine and human fibrinogen (SHULMAN, 1953; CASPARY and KEKWICK, 1957). By the sedimentation equilibrium method the molecular weight of bovine fibrinogen was estimated to 400000 (JOHNSON and MIHALYI, 1965). Light scattering techniques and osmotic pressure measurements have given different values. In solutions of purified fibrinogen, aggregates tend to form during storage, brought about by thrombin which tends to be present as a contaminant. Traces of plasmin are also often present producing degradation of the molecule. These facts, which were recognized only recently, explain the rather wide range of molecular weights reported in the literature.

Studies on the conformation of fibrinogen with various methods (optical rotary dispersion, low-angle X-ray diffraction) indicate that fibrinogen is a helical protein with 30–35% helix (HUSEBY and MURRAY, 1967b; STRYER et al., 1963). HALL and SLAYTER (1959) have presented a structural model for fibrinogen, based on electron microscopical evidence, suggesting that the fibrinogen molecule consists of a rod about 475 Å long with two terminal nodules about 60 Å in diameter and a central nodule about 50 Å. BANG (1964) has reported a length of 375 ± 40 Å and a width of 60 ± 10 Å.

The amino acid composition of fibrinogen has been studied by automated analysis (HENSCHEN and BLOMBÄCK, 1964; HUSEBY and MURRAY, 1967a). N-terminal analysis of bovine fibrinogen was first made by BAILEY et al. (1951) and extended studies of fibrinogen and fibrin from different species have been done by BLOMBÄCK and YAMASHINA (1958), using the method of EDMAN (1950). A common feature in the different fibrinogens is the occurrence of N-terminal tyrosine. The occurrence of the N-terminal tyrosine is a common feature also in mammalian fibrin. The other N-terminal amino acids found in fibrinogen disappear during conversion to fibrin showing that a limited proteolysis is involved in the fibrinogen-fibrin transformation.

Fibrinogen consists of two identical parts containing three different peptide chains, indicating that it exists as a dimer (CLEGG and BAILEY, 1962; HENSCHEN, 1964). The molecular weight of the subunits has been calculated to 180000 (CAPET-ANTONINI and GUINAND, 1967).

The N-terminal pattern of the three peptide chains, α ("A"), β (B) and γ, have been determined by BLOMBÄCK et al. (1966, 1972). They are linked together in a firm disulfide knot (BLOMBÄCK et al., 1968). Specific fibrinopeptides are released from the α ("A") and β (B) chains by thrombin whereas the γ chain remains unaltered.

The dominant role of the liver in the biosynthesis of fibrinogen and other clotting factors has been demonstrated with the aid of the isolated perfused rat liver (MILLER et al., 1964; MILLER and JOHN, 1970) and by immunofluorescent research of the hepatocytes (BARNHART et al., 1970). The fibrinogen level is therefore reduced in severe liver disease.

About 100 cases of afibrinogenemia and hypofibrinogenemia have been described. The heredity is autosomal recessive. In afibrinogenemia, the bleeding disorder may be mild, but more surprising is the observation of 3 cases who died from thromboembolism although they had serious hypofibrinogenemia (INGRAM et al., 1966).

In recent years about 30 cases with an abnormal fibrinogen molecule have been reported (see MÉNACHÉ, 1970). The inheritance is autosomal dominant. Some of these patients are asymptomatic, others have a moderate bleeding tendency and two cases were hospitalized because of thrombosis (JACKSON et al., 1965; EGEBERG, 1967). Differences in the amino acid composition with substitution of one amino acid with another have been demonstrated in such cases (BLOMBÄCK and BLOMBÄCK, 1970).

Factor II (Prothrombin)

BORDET and DELANGE (1914) removed prothrombin from plasma by adsorption with calcium phosphate. Various methods of adsorption and elution from anorganic precipitates have since been used for the isolation of prothrombin. Prothrombin of high quality from bovine plasma was first prepared by SEEGERS (1952, 1964). He obtained a protein which was homogeneous in the analytical ultracentrifuge and electrophoretically, and which could be converted rather quantitatively into thrombin. The purified prothrombin migrated with the α_2-globulin by electrophoresis and the isoelectric point was estimated to be pH 4.2. In spite of the homogeneity, these prothrombin preparations were later shown to contain factors VII, IX and X. By chromatography on DEAE-cellulose these factors could be removed. Human prothrombin has been prepared in many laboratories. The highest degree of purity was obtained by LANCHANTIN and FRIEDMAN (1963) and SHAPIRO and WAUGH (1966). Various combinations of adsorption, elution and chromatography have been used for purification such as DEAE-cellulose (MAGNUSSON, 1965c), Sephadex G-200 (MAGNUSSON, 1965b; TISHKOFF et al., 1968), gel filtration (MAMMEN and RAMIEN, 1962), Amberlite IRC-50 (MILLER, 1958).

The molecular weight for the bovine prothrombin complex has been calculated to 68000 and 68500 (LAMY and WAUGH, 1958; HARMISON et al., 1961), and the result for bovine prothrombin was reported to be 65500 by TISHKOFF et al. (1968). The molecular weight of human prothrombin was given as 68700 (LANCHANTIN et al., 1968). SEEGERS et al. (1967) found that the prothrombin complex contained 526 amino acid residues. On the basis of the amino acid composition the molecular weight was calculated to 58800. The weight of the carbohydrate content was assessed at approximately 8000 giving a molecular weight for the entire prothrombin complex of 66800. A total of 556–559 amino acid residues have been found for human prothrombin (LANCHANTIN et al., 1968). The N-terminal amino acid of bovine prothrombin is alanine (MAGNUS-SON, 1965a). Bovine prothrombin was found to contain carbohydrate in a total amount of 11.2–11.6% (SCHWICK and SCHULTZE, 1959; MAGNUSSON, 1965b). For human prothrombin a total carbohydrate content of about 10% has been found (LANCHANTIN et al., 1968).

Purified preparations of the prothrombin complex, which contain trace amounts of other clotting factors, are converted to thrombin spontaneously or in 25% sodium citrate solution. Prothrombin which is prepared completely free of other factors does not convert to thrombin, neither by thromboplastin, factor V and calcium, nor by sodium citrate (ALEXANDER, 1958).

Prothrombin is synthesized by the liver and vitamin K is required for normal production. The plasma level is reduced in vitamin K deficiency (obstructive jaundice, malabsorption) and by anticoagulant therapy with cumarin drugs. The synthesis is blocked in one of its last steps. Factors VII, IX and X are depressed simultaneously with prothrombin and a protein appears in the circulating blood, termed PIVKA (protein induced by vitamin K absence) which acts as a competitive inhibitor in the reaction sequence which activates factor X (HEMKER et al., 1968a, b). Immunological techniques have demonstrated an excess of immunologically detectable prothrombin over prothrombin activity (GANROT and NILÉHN, 1968; JOSSO et al., 1968), of factor X over factor X activity (PRYDZ and GLADHAUG, 1971), of factor IX over factor IX activity (LARRIEU and MEYER, 1970; VELTKAMP et al., 1971; LECHNER, 1972) and also of factor VII over factor VII activity (GOODNIGHT et al., 1971).

Only a few patients with hereditary prothrombin deficiency have been described (see KATTLOVE et al., 1970). In some of these cases the presence of an abnormal prothrombin molecule was demonstrated (SHAPIRO et al., 1969; KATTLOVE et al., 1970).

The heredity seems to be autosomal recessive, and the bleeding disorder is mild, probably because the lowest prothrombin activity in such patients has been about 10%. A complete lack of prothrombin activity would presumably be incompatible with life.

Factor III (Tissue Thromboplastin, Tissue Factor)

MORAWITZ (1905) found that aqueous extracts of tissues have a greater coagulant activity than the lipid, extracted by alcohol, which had been used by SCHMIDT (1892). HOWELL (1912) suggested that the active component of tissue extracts was a lipoprotein. CHARGAFF and co-workers (1944, 1945) prepared a purified lipoprotein activator from lung with a high activity and found a cephalin-like lipid component. By electron microscopy spherical particles with a diameter of 80–120 mμ and a calculated weight of 167 000 000 were found. Alcohol-ether extraction removed about 50%. About 45% was recovered as lipids containing 63% phospholipids consisting of 25% cephalin, 25% lecithin and 12% sphingomyelin. Tissue extracts have a heterogeneous composition and often contain a mixture of procoagulant and anticoagulant activity. Both the qualitative and quantitative composition vary with the source of material (brain, lung, placenta) and this largely determines their specificity and properties. HECHT et al. (1958) purified thromboplastin from rabbit brain and found that it was not a protein but a complex lipid consisting of sterol, glutamic acid, serine, ethanolamine and sphingosine. DEUTSCH et al. (1964) and HVATUM and PRYDZ (1969) however, found that brain thromboplastin was a lipoprotein. They separated the protein component from the lipid and found that the latter had procoagulant effect, like cephalin, whereas the protein was inactive. By recombination the full thromboplastic activity could be restored. DEUTSCH et al. found that the lipid fraction contained phosphatidyl ethanolamine, phosphatidyl choline, phosphatidyl serine, phosphatidyl inositol, lysophosphatidyl ethanolamine and sphingomyelin.

The preparations used by different researchers have varied greatly because they have been prepared differently, by extraction of either fresh tissue or acetone dried tissue, and by using brain, lung or placenta, from rabbits, human or other sources. The qualitative differences in activity have created difficulties in the interpretation of the one-stage prothrombin time test. Differences in results are very great by testing of the same plasma from a patient on anticoagulant treatment with different thromboplastins. One of the reasons is that some thromboplastins are insensitive to depressions of factor VII, or X, or both, in the test-plasma, because of contaminations by serum or coagulation intermediates from the organ and for used preparation. Another reason is that there are differences between thromboplastins in their sensitivity to coagulation inhibitors, and particularly to the PIVKA inhibitor which occurs during anticoagulant therapy (HEMKER et al., 1963, 1964, 1968a, h) The standardization of thromboplastins therefore is an urgent issue which has been studied by an international cooperation of expert laboratories (BIGGS, 1965; BIGGS and DENSON, 1966, 1967; BIGGS, 1970).

Factor IV (Calcium)

The requirement of calcium for coagulation has been recognized since ARTHUS and PAGES (1890) inhibited coagulation by the precipitation of calcium and reversed the effect by adding calcium. QUICK (1940) observed that the prevention of coagulation required greater amounts of oxalate than calculated to be necessary for neutralizing ionized calcium and suggested that the calcium which is active in coagulation, is bound to prothrombin. Calcium is required for all coagulation reactions, except for the initial contact activation of factors XII and XI and for the action of thrombin on fibrinogen, but it accelerates this latter reaction. It is required however for the activation of factor XIII by thrombin and thereby for the production of a normal, stable fibrin clot (LAKI and LORAND, 1948).

Factor V (Proaccelerin, Labile Factor)

The first concentrates of factor V were prepared from adsorbed plasma by ammonium sulphate (33–50% saturation) and isoelectric fractionation (OWREN, 1947a, 1948; WARE and SEEGERS, 1948). Preparations of higher purity have been obtained by chromatographic techniques. ESNOUF and JOBIN (1967) have reported a 4000-fold purification and BARTON and HANAHAN (1967) a purification 3 500–8 500-fold. Factor V in plasma is not adsorbed by barium sulphate but the greater part can be adsorbed by soybean lecithin (SEAMAN and OWREN, 1956).

AOKI et al. (1963) have reported that SH-groups are essential for its activity. The molecular weight of bovine factor V has been reported differently as 97000 (AOKI et al., 1963), 180000 (HUSSAIN and NEWCOMB, 1963), 290000 (ESNOUF and JOBIN, 1967) and 400000 (PAPAHADJOPOULOS et al., 1964).

Factor V is activated both by thrombin and by RUSSELL'S viper venom (WARE et al., 1947; HJORT, 1957; BERGSAGEL and NOCKOLDS, 1965). The activation capacity varies between species. Whereas human factor V activity increases 20 times, bovine factor V increases 40 times by thrombin, and the activated factor becomes adsorbable onto barium sulphate and has a different isoelectric point (STORMORKEN, 1957). KAHN and HEMKER (1972) estimated the molecular weight of unactivated human factor V by gel filtration to 410000. After activation with thrombin or RUSSELL'S viper venom the molecular weight was 110000, suggesting the dissociation of a tetramer. The approximate molecular weight for unactivated and activated bovine factor V was 400000 and 195000, suggesting the dissociation of a dimer.

When human factor V is activated by thrombin a rapid deterioration of the activity occurs in a few hours (Lewis and Ware, 1953; Cox et al., 1956; Stormorken, 1957). The activation by thrombin followed by inactivation is probably the reason for the absence of factor V in human serum. Bovine factor V is less easily destroyed by thrombin.

Unactivated factor V is unstable in decalcified plasma and disappears during storage presumably by denaturation. The velocity of denaturation depends upon pH and the concentration and nature of the anticoagulant, and stability is highly improved by the presence of calcium ions (Stormorken, 1957; Blombäck and Blombäck, 1963). Preparations of factor V can be freeze-dried with moderate loss of activity. In solutions in the presence of calcium it keeps well for a few months at $-25°C$. The inactivation is markedly reduced by glycerol.

Since the first case of factor V deficiency was discovered about 50 cases have been described. The mode of inheritance seems to be autosomal recessive. Both homocygotes and heterocygotes have been recorded but the bleeding tendency in both instances is rather mild. Also in this deficiency the occurrence of thrombosis has been described (Miller, 1965). A familiar abnormality with increased factor V and concomitant thrombosing tendency has also been reported (Gaston, 1966). Whether an abnormal factor V molecule is present in the congenital deficiencies is not yet known.

Factor V decreases in severe liver diseases (Owren, 1949). At least part of factor V therefore seems to be synthesized in the liver.

Factor VII (Proconvertin)

The existence of additional factors of factor V, thromboplastin and calcium, in the extrinsic clotting system for prothrombin conversion has been suggested by several workers (Owren, 1947a, b, 1951; Owen et al., 1948, 1951; De Vries et al., 1949; Alexander et al., 1949). The factor was first termed co-factor V by Owren (1947b) and later proconvertin (Owren, 1951, 1952), serum pro-thrombin conversion accelerator (SPCA) by Alexander et al. (1949) and factor VII by Koller et al. (1951, 1952). Factor VII is present in serum. The clot promoting activity of serum was observed already by Bordet and Gengou in 1904.

The isolation and purification of factor VII from plasma or serum have been attempted by several investigators (Deutsch and Schaden, 1953; Duckert et al., 1953; Alexander, 1958; Tishkoff et al., 1960; Prydz, 1964; Deutsch et al., 1966; Shaw et al., 1966; Högenauer et al., 1968 and others). It is adsorbed and eluted similarly to prothrombin and factors IX and X.

It has also been isolated from the activated purified prothrombin complex (MAMMEN et al., 1960a, b, SEEGERS et al., 1962). It has been obtained free of factors IX and X by chromatography on DEAE-sephadex (PRYDZ, 1964; GLADHAUG and PRYDZ, 1970). They estimated the molecular weight to about 63000 and 48000 for the plasma and serum factor VII respectively. This indicates proteolytic cleavage during activation. No caseinolytic or esterolytic activity of the preparations were found. Concentrates, containing also factor II, IX and X, for treatment of these deficiencies have been prepared (SOULIER et al., 1969).

The first cases of congenital deficiency of factor VII were studied by ALEXANDER et al. (1951), OWREN (1952a, b) and AAS (1952). About sixty cases have since been reported. The inheritance is autosomal recessive. The homocygotes are mild bleeders, whereas the heterocygotes have no clinical symptoms. Whether patients with factor VII deficiency also are a polymorphic group with various defective forms of factor VII has not yet been decided. The one-stage prothrombin time in factor VII deficiency is prolonged but if tissue thromboplastin is substituted by RUSSELL's viper venom and cephalin, the clotting time will be the same as for normal plasma. This test is useful for the differentiation of factor VII and factor X deficiency.

Factor VII is synthesized in the liver in the presence of vitamin K. The plasma level of factor VII is reduced by treatment with cumarin drugs, in K-avitaminosis, in liver diseases and low levels are seen in the newborn. A prestage of factor VII occurs in the circulation during anticoagulant treatment and K-avitaminosis as already mentioned (p. 7).

Factor VIII (Antihaemophilic Factor A)

The name haemophilia was first used by HOPFF in 1828. It was established as a specific clinical entity by BULLOCK and FILDES in 1911. The abnormal clotting was probably first observed by LISTON in 1839. WRIGHT in 1893 described the prolonged clotting time. ADDIS (1911) found that addition of small amounts of normal plasma corrected the clotting time. The deficiency of a plasma factor was definitely established by PATEK and TAYLOR in 1937 and it was called "antihaemophilic globulin" (AHG). BRINKHOUS (1939) observed that tissue thromboplastin would normalize the thrombin formation in haemophilic plasma. His observation not only confirmed the lack of a plasma factor but demonstrated a pathway of coagulation which was independent of tissue thromboplastin.

Until 1947 haemophilia was regarded as an entity. PAVLOVSKY then observed that blood from two patients with typical clinical haemophilia was mutually corrective both in vitro and in vivo. In 1950 KOLLER et al. described a family with the clinical picture of classical haemophilia, but with a different deficiency. Studies by Biggs et al. (1952) and Aggeler et al. (1952) definitely established the second type of haemophilia caused by the lack of a plasma factor, different from factor VIII. Biggs and co-workers termed the factor Christmas factor after their patient, and Aggeler et al. called it "plasma thromboplastin component" (PTC). The factor has later been assigned number IX. Factor VIII deficiency is now usually termed haemophilia A and factor IX deficiency haemophilia B.

Haemophilia A is characterized by deficient or nearly absent acticity of factor VIII and it was long believed that this was caused by an inability to synthesize the factor VIII protein. Investigations with the use of human circulating antibodies, which had appeared in haemophilia A patients after transfusions, revealed the existence of two types of patients. One type was lacking both the clotting activity and the antigenic activity of factor VIII and was termed haemophilia A—, whereas the other type, being about 10–15 % of the patients, had a low clotting activity but a high amount of cross-reacting material. This type was termed haemophilia A+ (Hoyer and Breckenridge, 1968; Denson et al., 1969; Feinstein et al., 1969). Recent studies, using precipitating (Zimmerman et al., 1971) or haemagglutinating (Stites et al., 1971) rabbit antibodies to highly purified factor VIII revealed that the majority if not all cases of haemophilia A have a normal amount of immunoreactive protein (Meyer et al., 1972).

In von Willebrand's disease, which is also characterized by a low factor VIII activity in plasma, was found a low level of cross-reacting material in plasma corresponding to the biological activity (Zimmerman et al., 1971; Bouma et al., 1972). It seems therefore, that the low factor VIII activity invon Willebrand's disease is caused by an inability to synthesize the factor VIII protein, whereas patients with haemophilia A synthesize an abnormal factor VIII molecule which has a low or nearly absent biological activity.

The non-functional factor VIII protein is concentrated in Cohn's fraction I and by cryoprecipitation similarly to the normal factor VIII. When normal factor VIII is treated with thrombin the activity first increases and then is quickly lost (Rapaport et al., 1963). The antigenic properties however, remain after thrombin inactivation (Hoyer and Breckenridge, 1968), illustrating that antigenic and procoagulant activities can be separated. Further studies with antibodies of human and rabbit origin have suggested that structurally different abnormal factor VIII proteins might exist, with different coagulant activities. This would explain the variation from severe to moderate forms of haemophilia A. Similarly structurally defective forms have been found for fibrinogen and factors IX and X and it is very likely that it will be found for the other clotting factors as well.

Purification of factor VIII has been tried in many ways, primarily in order to obtain clinically useful preparations. Factor VIII follows fibrinogen in most procedures and similar methods, as for the concentration of fibrinogen, have been used, such as isoelectric precipitation, salting out procedures, fractionation with ethanol or ether and cryoprecipitation. The cryoprecipitation technique, which was introduced by Pool et al. in 1964, gives a 10 to 20-fold purification over plasma (Newman et al., 1971). These preparations have been very useful clinically, although they have a high content of fibrinogen. Separation of factor VIII from fibrinogen, and further purification, has been obtained by various procedures (Blombäck, 1958; Blombäck et al., 1958; Simonetti et al., 1961; Johnson et al., 1967; Brinkhous et al., 1968; van Mourik and Mochtar, 1970). Purification up to 10000 times over plasma was reported by Johnson et al. (1967).

The separation of factor VIII from fibrinogen resulted in decreased stability and a change in its precipitation properties (Barrow et al., 1966). Instability was also observed when factor VIII was isolated from plasma of a patient with congenital afibrinogenemia (Gobbi, 1960). The extreme lability of factor VIII has made it difficult to analyse its physico-chemical characteristics. Shulman et al. (1960) calculated the molecular weight of bovine factor VIII to 196000. Seegers et al. (1957) determined the isoelectric to pH 6.4 and Lewis et al. (1958) found that factor VIII was associated with α_1 and α_2 globulins. The highly purified preparation of Johnson et al. (1967) still showed two components by acrylamide-gel electrophoresis. This preparation was surprisingly stable upon storage. Amino acid composition and terminal amino acids for factor VIII have not been reported.

Factor VIII is produced in the spleen (Pool, 1966), in the liver and the kidneys (Barrow and Graham, 1968). The spleen is a rich source of factor VIII and transplantation of a normal spleen to a haemophilic dog increased the plasma level (Norman et al., 1968). Transfusions of human spleen suspensions to patients with haemophilia A have also been tried (Desai, 1969).

Haemophilia is transmitted as a sex-linked recessive character and only a few cases have been recorded in females. About $1/3$ of the cases are without a positive family history. The incidence of haemophilia A is about one in 50000. In genetic counselling it is important to diagnose the carriers. This has been successful to a certain extent by measurement of factor VIII activity, preferably in combination with a determination of prothrombin consumption and/or factor V during coagulation (Miller and Siggerud, 1964; Nilsson et al., 1959; Veltkamp et al., 1968; Owren, 1950). Further studies with the immunological technique will probably improve the diagnosis of carriers.

A familial increase in factor VIII associated with an increased thrombotic tendency has been reported (Penick et al., 1965).

Factor IX (Antihaemophilic Factor B)

Factor IX is adsorbed and eluted in the same manner as prothrombin and factors VII and X. Concentrates of the four factors, prepared by adsorption and elution, have been used successfully in haemophilia B for preventing bleeding following major surgery and dental extraction (Soulier et al., 1969). The purity of factor IX preparations isolated from plasma and serum have so far been unsatisfactory for physicochemical characterization. By preparation from the purified prothrombin complex a product of higher purity was obtained (Mammen et al., 1960a). The N-terminal amino acid was found to be proline and the C-terminal amino acid was tyrosine. The purified factor IX preparation contained 20% carbohydrates and the molecular weight for the entire glucoprotein was calculated to 51 200 (Mammen, 1964). Aggeler et al. (1954) found that factor IX migrated in the β_2 region. Harmison and Seegers (1962) calculated a molecular weight of 49 900 using the Svedberg equation.

Shortly after the discovery that the clinical picture of haemophilia included two distinct entities, haemophilia A and B, the latter was found to exist in two types, one with and one without factor IX-like material in plasma (Fantl and Sawers, 1956; Fantl et al., 1956). Cases with a defective form of factor IX in plasma were described by Hougie and Twomey (1967). These cases had a prolonged prothrombin time with ox brain thromboplastin indicating that the abnormal factor IX inhibited the formation of prothrombinase. They termed this type haemophilia Bm. Roberts et al. (1968), by means of a specific factor IX antibody preparation, found that about 10% of the Haemophilia B patients had an abnormal factor IX molecule. The remaining 90% was assumed to have a deficient synthesis. Further studies showed however, that many of these latter patients also had a normal amount of antigenic factor XI material in plasma. This material had a low biological activity, but was not inhibitory (Pfueller et al., 1969; Veltkamp et al., 1968, 1970; Larrieu and Meyer, 1971; Denson et al., 1968). This form was termed haemophilia B+ and patients without demonstrable factor IX antigen was termed haemophilia B—.

The differentiation of haemophilia Bm from the other types is easy by using the Thrombotest method because it is sensitive to the inhibitor. Haemophilia Bm gives a prolonged Thrombotest time in contrast to the normal Thrombotest time in the other two forms (Larrieu and Meyer, 1971). If the abnormal factor IX in haemophilia Bm is removed by the addition of factor IX antibody the Thrombotest time and the one-stage prothrombin time with ox brain thromboplastin becomes normal (Denson, 1971). Denson confirmed the two types of haemophilia B, with presence and absence of a substance in plasma which was antigenically similar to factor IX. There is still a possibility that haemophilia B— might have an abnormal factor IX molecule which is structurally altered in such a way that it has not yet been detected by immunological techniques. There might in fact exist a broad spectrum of abnormal factors in both haemophilia A and B.

Factor IX is synthesized by the liver in the presence of vitamin K. It is reduced in K-avitaminosis and during anticoagulant therapy. In these conditions a precursor molecule of factor IX, having the antigenic properties but defective function, appears in the circulation (see p. 7, 35). Congenital factor IX deficiency is less frequent than factor VIII deficiency, the proportion being about 1:5. It is inherited as a sex-linked recessive character similarly to haemophilia A. It is possible that the mode of inheritance is different for certain of the variants of both haemophilia A and B, but this problem has not yet been thoroughly studied.

Factor X (Stuart-Prower-Factor)

Studies of dicumarol plasma by DUCKERT et al. (1954, 1955) suggested the existence of an adsorbable clotting factor in serum in addition to factor VII. They named it factor X. TELFER et al. (1956), HOUGIE (1956) and HOUGIE et al. (1957) studied patients with haemorrhagic states who had a prolonged one-stage prothrombin time, but in contrast to factor VII deficiency also had a prolonged clotting time with RUSSELL's viper venom and cephalin. Serum was abnormal in the thromboplastin generation test and the lacking factor therefore could neither be factor VII, nor factors V or IX. It was named Prower factor and Stuart factor after their patients. It was soon realised that all these factors were the same.

Factor X has been isolated from both plasma and serum by several investigators. Methods of isolation of bovine factor X have been reported by ESNOUF and WILLIAMS (1962), PAPAHADJOPOULOS et al. (1964), LECHNER and DEUTSCH (1965) and JACKSON et al. (1968). Human factor X has been isolated by KAHN and BOURGAIN (1965) and GLADHAUG and PRYDZ (1970). SEEGERS et al. (1964b) and SEEGERS and MARCINIAK (1965) isolated factor X (autoprothrombin III) from purified prothrombin complex activation mixtures.

ESNOUF and WILLIAMS (1962) reported a 4000-fold purification and JACKSON et al. (1968) stated that the final product represented a 16000-fold purification, with a yield of 1 mg per litre of plasma. Both materials were homogeneous in the ultracentrifuge.

GLADHAUG and PRYDZ (1970) obtained factor X preparations, purified about 10000-fold, from human serum and free from the coagulation factors II, V, VII and IX. Rabbits injected with these preparations produced antibodies against factor X. The antibody neutralized completely both non-activated factor X and factor X activated by RUSSELL's viper venom in the presence of cephalin and calcium.

Purified factor X migrates in the area of the α_1-globulin. The molecular weight of bovine factor X has been calculated to different values, 84800–87000 by Esnouf and Williams (1962), 52600 by Seegers et al. (1969), 55000 by Jackson and Hanahan (1968) and 74000 by Murano (1968). All preparations of factor X could be activated by Russell's viper venom, lipid and calcium. Some preparations activated in 25% sodium citrate and also spontaneously.

Macfarlane (1961) and Williams and Esnouf (1962) first discovered that factor X was the substrate for Russell's viper venom and they purified the venom and studied the interaction.

Williams and Esnouf calculated the molecular weight of Xa to 36000. Seegers et al. (1963, 1966) have also isolated bovine activated factor X (autoprothrombin c) and calculated the molecular weight to 21900 (Seegers et al., 1969). They suggested the possibility that factor X is a dimer of two active factor X molecules.

Esnouf and Williams (1962) determined the N-terminal amino acids of factor X as alanine and glycine, whereas Högenauer et al. (1968) found glycine and serine and Murano (1968) glycine, alanine and serine. Murano found the same for activated factor X, whereas Esnouf and Williams found alanine, glycine and either leucine or isoleucine. Activated factor X had esterolytic activity (Esnouf and Williams, 1962). Activated factor X is inactivated by antithrombin (Seegers et al., 1964a; Egeberg, 1965; Biggs et al., 1970; Yin and Wessler, 1970).

Several cases of congenital factor X deficiency have been reported and the bleeding tendency is rather severe. The inheritance of factor X deficiency is autosomal, incompletely recessive, but highly penetrant (Graham et al., 1957). The heterozygotes have about 20–50% factor X and might therefore be identified in most instances. It has recently been demonstrated that a significant number of patients with factor X deficiency have a defective factor X molecule, with the same antigenic properties as the normal factor X but without biological activity (Shapiro et al., 1970; Denson et al., 1970; Denson, 1971). Studies by Denson (1971) suggest that the abnormal factor X molecule differs in different unrelated families and that a spectrum of abnormal factor X molecules may arise as a result of genetic mutations. All the patients investigated had some functional activity of the abnormal factor X molecule and it is suggested that a complete lack of functional factor X activity might be lethal.

Factor X is synthesized in the liver in the presence of vitamin K. It is reduced in vitamin K deficiency and during anticoagulant therapy with cumarin drugs. Prydz and Gladhaug (1971) found, with the use of specific antibody against factor X, that plasma from patients treated with cumarin drugs had all essentially the same capacity (80–100%) as normal plasma to adsorb the factor X antibodies in spite of the low factor X activities. The existence of an antigenically similar but inactive precursor of factor X in anticoagulant therapy suggests an inhibition of the synthesis of factor X at a late stage, when the antigenic structures, cross-reacting with those of plasma factor X, have already been formed.

Separate depression of factor X has been observed in amyloidosis (KORSAN-BENGTSEN et al., 1962; MÉNACHÉ and BOIVIN, 1962; PECHET and KASTRUL, 1964).

Factor XI (Plasma Thromboplastin Antecendent)

The reason why blood clots more rapidly in a glass tube than in a similar tube coated with paraffin, as observed by BORDET and GENGOU in 1903, remained an unsolved problem until the discovery of the contact factors, now known as factors XII and XI. ROSENTHAL et al. (1953, 1955) described a haemorrhagic state which differed both in its mode of inheritance and in the nature of the defect from haemophilia A and B. The deficient factor was called plasma thromboplastin antecedent (PTA) and has later been assigned number XI.

Factor XI is present in both plasma and serum. When it is adsorbed from plasma together with factor XII, by celite, glass, kaolin or $BaCO_3$, it becomes activated. Factors XII and XI probably form a complex on the surface of the adsorbent and the product can be eluted by 1.7 M sodium chloride solution, This product has been referred to both as "the activation product" and as activated factor XI. Partial separation of factor XI and XII from plasma was achieved by adsorption with low concentrations of celite (SOULIER and PROU-WARTELLE, 1960). RATNOFF et al. (1961) purified factor XI 10–20 fold. from factor XII deficient plasma. Purified preparations, up to 150-fold have been obtained from serum (KINGDON et al., 1964). The factor XI preparations showed esterolytic activity, but physico-chemical data have not been reported because of low purity. The site of biosynthesis is unknown.

More than 200 cases of congenital factor XI deficiency have been reported, indicating that it is more frequent than other congenital deficiencies, except haemophilia. It is inherited as an autosomal dominant or codominant trait (RAPAPORT et al., 1961). The bleeding tendency is milder than in the haemophilias and haemarthroses are rare. Some patients have no abnormal bleeding (EGEBERG, 1962).

Factor XII (Hageman Factor)

Factor XII was discovered by RATNOFF and COLOPY in 1955. The disorder was called HAGEMAN trait and the deficient factor HAGEMAN factor, named after the patient. Mr. HAGEMAN's clotting time was very prolonged, but he had no bleeding tendency, a finding which has puzzled the coagulationists ever since.

Concentrates of factor XII have been prepared by various procedures of adsorption and elution, ethanol precipitation, electrophoresis and column chromatography. Products up to 5000-fold purified over plasma have been isolated (RATNOFF and DAVIE, 1962a, b; SCHOENMAKERS et al., 1963). SCHOEN-MAKERS et al. (1965) obtained preparations of high specific activity which showed homogeneity in the analytical ultracentrifuge and a single boundary by disc electrophoresis. A molecular weight of 82000 was calculated for the activated form of factor XII. A recent and extensive study by GRAMMENS et al. (1971) of a purified preparation of bovine factor XII, which appeared homogeneous in all aspects, had a calculated molecular weight of 142000 with close agreement by different methods. SCHOENMAKERS et al. (1965) found that factor XII had a carbohydrate content of nearly 14% with 4.4% sialic acid and assessed factor XII to be a sialo-glucoprotein. Removal of sialic acid did not affect the coagulant activity. GRAMMENS et al. however, found a carbohydrate content of only 3% and sialic acid of 0.35%. By the EDMAN procedure, they found 2 N-terminal groups, glycine and valine. The isoelectric point was determined at pH 7.9 by electrofocusing technique. Factor XII migrated with γ-globulin. It was a powerful immunogen and antibodies against factor XII were rapidly elicited by a single injection of μg quantities into rabbits.

NIEWIAROWSKI et al. (1962) and SCHOENMAKERS et al. (1965) found that factor XII had esterase activity. SHERRY et al. (1966) assumed however, that the esterolytic activity probably was not due to factor XII itself, but to another enzyme which was activated by factor XII, presumably kallikrein.

About 100 cases with the HAGEMAN trait have been reported. The incidence is certainly much higher than expressed by this figure because there is no bleeding tendency, although EGEBERG (1970) has described a family with mild bleeding tendency. Two cases with factor XII deficiency developed myocardial infarction (GLUECK and ROEHLL, 1966; HOAK et al., 1966) and Mr. HAGEMAN himself died from thromboembolism (RATNOFF et al., 1968).

The site of synthesis is unknown, but the reduction of the plasma level in liver diseases, which has been reported by JÜRGENS (1962) might indicate that at least part of the HAGEMAN factor is synthesized in the liver.

Factor XIII (Fibrinase)

In 1948 LAKI and LORAND obtained evidence that plasma and serum contain a factor which is necessary for the formation of a normal clot. It was named the "fibrin stabilizing factor", and has been assigned number XIII. The first case of factor XIII deficiency was described by DUCKERT et al. (1961).

Factor XIII has been purified from human plasma by ammonium sulphate fractionation and chromatography on DEAE cellulose (LOEWY et al., 1961)

or Bio-Gel P200 (LORAND and KONISHI, 1966). Products up to 8000-fold purification have been prepared with a calculated molecular weight of 35000.
The molecule with a calculated molecular weight of 110000, under certain
conditions dissociated into subunits suggesting three units (LOEWY et al., 1961).
The structure has been studied by SCHWARTZ et al. (1971).

Factor XIII is a precursor which is activated by thrombin into an active
enzyme which acts as a transpeptidase (see p. 30).

About 30 cases of factor XIII deficiency have been reported (see EGE-
BERG, 1968; WALLS and LOSOWSKY, 1968). Factor XIII deficiency is characterized by repeated prolonged bleedings from injuries, usually starting after
24—36 hours, and most typically, delayed wound healing and formation of
extensive scar tissue. The inheritance is autosomal, incompletely recessive.

Acquired low levels of factor XIII have been reported in certain haemato-
logical disorders (NUSSBAUM and MORSE, 1964; GERHOLD et al., 1966) and
in liver diseases, particularly in advanced cirrhosis (OTTAVIANI et al., 1965;
WALLS and LOSOWSKY, 1968). The reduction in chronic liver disease indicates
that the liver takes part in its synthesis.

Lipid Procoagulants ("Cephalin")

The lipid procoagulants have not been assigned any Roman numeral. Their
coagulant effect has been recognized since SCHMIDT's (1892) experiments with
alcoholic extracts of tissues. The activity was thermo-stable, and BORDET and
DELANGE (1913) suggested that the active substance was of the lecithin group.
Lipid procoagulants have been obtained from platelets (platelet factor 3) and
from lung and brain thromboplastins by alcohol-ether extraction. All extracts
contain cholesterol, various phospholipids and sphingomyelin and much research has been devoted to finding the exact nature of the procoagulant
activity as discussed in the monograph by HECHT (1965). POOLE and ROBINSON
(1956) found activity in the phosphatidyl ethanolamine fractions of brain and
egg phosphatides and also demonstrated activity for three synthetic ethanolamine preparations. BIGGS and BIDWELL (1957) confirmed the activity in
phosphatidyl ethanolamine fractions, but found no correlation between the
ethanolamine content and the degree of activity. Purified preparations were
less active than the crude cephalin fraction. O'BRIEN (1957) showed that phosphatidyl serine had just as high activity as phosphatidyl ethanolamine. Purified
cholesterol had no effect and sphingosine (which is a part of sphingomyelin)
had anticoagulant activity. The active component therefore is a phospholipid.
Several conflicting conclusions have been reported concerning the effect of
the various phospholipids, probably varying with the purity of the preparations and with the test system used. Phosphatidyl serine was found to act
as a substitute for platelet factor 3 (MARCUS et al., 1962), but as an inhibitor

in the thromboplastin generation test (Turner and Silver, 1963) and as an anticoagulant by infusion (Mustard et al., 1962). The highest coagulant activity has been obtained by a combination of phospholipids such as of phosphatidyl ethanolamine and phosphatidyl choline (Rapport, 1956), or phosphatidyl choline and phosphatidyl serine (Troup et al., 1960) or phosphatidyl serine and phosphatidyl ethanolamine (Hecht and Slotta, 1962). Further studies have revealed that the activity of phospholipids is related to the surface charge of the lipid micelles (Bangham, 1961; Papahadjopoulos et al., 1962; Silver et al., 1963). A negative zeta-potential seems to be required for the coagulant effect. This might explain the findings that differences in activity also depend on the colloidal state of the phospholipid particles, pH, the degree of unsaturation and the localization of the fatty acid constituents on the phospholipids.

Certain clotting factors are adsorbed onto the surface of the phospholipid micelles and interactions and enzyme activations take place on the surface (wide infra).

Suggested New Clotting Factors

Several new clotting factors, in addition to those mentioned, have been suggested, but have not been accepted by the International Committee as real or original entities with sufficient scientific evidence for their existence to be assigned a Roman numeral.

The Prephase Accelerator (PPA) (Duckert, 1961). The factor accelerated the contact phase of the intrinsic clotting system. It required factors XII, XI and IX for its formation. It did not correct the haemophilia B defect in the thromboplastin generation test. PPA is presumably a reaction product formed during coagulation and remaining in serum, and not a genuine clotting factor.

Thromboplastin Generation Accelerator (TGA) (Spittel et al., 1960; Pascuzzi et al., 1961). TGA was described as an accelerator of the early stages of coagulation and was found to be in excess amounts in certain thrombotic states. It was also present in plasma from patients receiving dicumarol, but was absent in human serum. It was non-dialysable, was associated with β-globulin on paper electrophoresis and was not adsorbed by barium sulphate. Its exact nature and function has not been reported.

Labile Serum Factor (Connor et al., 1961). This activity developed by storage of serum from some patients with liver disease, polycythemia vera and in dicumarol treated patients. Its nature has not been elucidated.

Thorium Vulnerable Factor (TVF) (Alexander and Colman, 1959). Thorium chloride, thorotrast and salts of related rare earth elements selectively and progressively inactivated a specific serum constituent which was essential for prothrombinase formation. The inactivation was reversible by citrate or oxalate. The factor was distinct from factors VII, IX, X, XI and XII, was adsorbable by $BaSO_4$, was reduced in patients on dicumarol, in liver diseases and in the neo-natal period.

Vasculokinase (VK) (MURRAY, 1961). A clotting factor which was extracted from aorta and which acted directly upon fibrinogen, but was different from thrombin. Studies indicated that its chemical activity probably was similar to that of staphylococcal coagulase.

Fraction P from Urine (CALDWELL et al., 1963). A procoagulant material prepared from urine which normalized the clotting time of haemophilia A plasma.

Serum Thrombotic Accelerator (STA) (WESSLER, 1960). STA is a potent thrombogenic activity found in serum. It causes stasis thrombosis by infusion. Probably it is an intermediate product which results from the activation of factors XII and XI.

The relation of the Fletcher factor to factors XI and XII is still not settled (HATHAWAY and ALSEVER, 1970), and the same also applies to the Hageman cofactor (OGSTON et al., 1971). There is also evidence that the complement system might be involved in the initial phase of coagulation (ZIMMERMAN et al., 1971). It is conjecturable whether a genuine clotting factor is lacking in the "Dynia" clotting abnormality (PECHET et al., 1967).

The Antibleeding Factor in von Willebrand's Disease

Von Willebrand's disease is characterized by prolonged bleeding time, reduced factor VIII activity, moderate disturbance of platelet adhesiveness and a response of factor VIII to transfusion of blood or plasma fractions which is different from that of haemophilia A. SHULMAN et al. (1956) and NILSSON et al. (1957) observed that transfusions of Cohn's fraction I, containing fibrinogen, factor VIII and other proteins, shortened the bleeding time and restored the factor VIII level. The prolonged bleeding time was also restored by transfusion of a fibrinogen preparation which was devoid of factor VIII activity after storage, indicating the presence in this preparation of an anti-bleeding factor. The factor was present in fresh haemophilia A plasma (NILSSON and BLOMBÄCK, 1959). Recent studies with the use of factor VIII antibodies have demonstrated a parallelism between factor VIII-protein and the biological activity in von Willebrand's disease indicating a defective synthesis of a normal factor VIII, whereas most if not all cases of haemophilia A are caused by the synthesis of a non-functional factor VIII molecule which has retained the antigenic properties of the normal factor VIII (see p. 12). It is possible therefore that the anti-bleeding effect is produced by factor VIII, both by the normal factor and the non-functional factor VIII protein in haemophilia A and in the fibrinogen preparation, which was used for transfusion in Nilsson et al.'s experiment, and which had lost its biologic factor VIII activity on storage. The paradoxical-rise in factor VIII activity in von Willebrand's patients in response to normal or haemophilic plasma transfusions (Nilsson and Blombäck, 1963; Biggs and Matthews, 1963; Lewis, 1964; Stormorken, 1967) has been the subject of many speculations, but there seems to be no reason for including the anti-bleeding factor among the coagulation factors taking part in the clotting mechanism. As it seems, the anti-bleeding activity is another function of the factor VIII molecule, different from its clot promoting activity.

The Interaction of the Clotting Factors

The Intrinsic Pathway

Several theories have been proposed in recent years to explain the interactions of the various clotting factors. In 1964 the "cascade" or "waterfall" theory was suggested for the enzymatic reaction mechanism of the intrinsic pathway of coagulation (Davie and Ratnoff, 1964; Macfarlane, 1964). According to this theory the first factor (factor XII) is activated by surface contact and thereby triggers a series of six consecutive activations, in which one activated factor acts as an enzyme and converts the next one, serving as substrate, into an active enzyme. Evidence which has accumulated in later years has made certain modifications of this theory necessary. A working hypothesis for the coagulation mechanism according to present knowledge is illustrated in Fig. 1. Three of the sequences in the cascade theory have been substituted with the formation of complexes of clotting factors with phospholipid and calcium. The concept can be summarized as follows: When a suitable surface is present, factor XII and factor XI interact and form an activator complex which in the presence of Ca^{++} activates factor IX. In the presence of phospholipid (platelet factor 3) and Ca^{++} activated factor IX and factor VIII form a complex which activates factor X. There is evidence that a preactivation of factor VIII by thrombin is necessary for its participation in this reaction (Rapaport et al., 1965). In the extrinsic pathway, tissue thromboplastin interacts with factor VII in the presence of Ca^{++} forming a complex which activates factor X. Factor X consequently joins the intrinsic and extrinsic clotting systems into a common pathway, in which activated factor X in the presence of factor V, phospholipid and Ca^{++} form a complex which activates prothrombin to thrombin. In the last main sequence, thrombin splits fibrinogen into fibrino-peptides and fibrin monomers which subsequently polymerize. Stabilization to insoluble fibrin follows by an effect of factor XIII and Ca^{++}. Pre-activation of factor XIII by thrombin is required for this reaction. In the following we shall discuss the mechanisms of these various reactions.

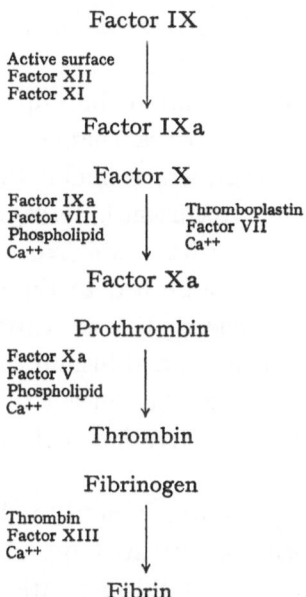

Fig. 1. Simplified scheme of the coagulation mechanism

The Contact Activation

LISTER (1863) observed that coagulation of blood in a glass tube started at the contact area with the glass. Studies of the contact phenomenon were greatly facilitated by the observation of JACQUES et al. (1946), that silicon-coating largely prevents the clot promoting effect of glass. The breakthrough in the study of the contact phenomenon came with the discoveries of factors XII and XI. Patients with deficiencies of either of these factors have an abnormally long clotting time if the test is carried out in a glass tube, but by clotting in silicon-coated tubes, the clotting times do not differ much from the clotting time of normal blood. Several workers have established that factors XII and XI are involved in contact activation (MARGOLIS, 1957; RATNOFF and ROSENBLUM, 1958; BIGGS et al., 1958; VROMAN, 1958; HARDISTY and MARGOLIS, 1959; SOULIER et al., 1959; WAALER, 1959 and others). Many clot promoting substances have been studied such a glass, celite, kaolin, $BaCO_3$ and asbestos, and recently also some substances of possible physiological relevance such as collagen, long chain fatty acids, and sodium urate. Only insoluble activating substances were known until RATNOFF and CRUM (1964) described activation of the Hageman factor by solutions of ellagic acid in concentrations as low as 10^{-8} M.

The wettability of a surface has often been suggested as an essential requirement for the activating property (VROMAN, 1967), but this opinion has not been supported by experimental findings. Certain non-wettable surfaces can activate coagulation and a colloidin-coated glass, which is just as wettable as glass, delays coagulation. The normal vascular endothelium, which prevents activation, is also wettable (Ratnoff, 1966).

A negative surface charge seems to be important, although there is no correlation between the total surface charge (zeta-potential) and the clot promoting effect. All substances, which inhibit the contact phase of coagulation, are positively charged (cytochrome c, lysozyme, ribonuclease, protamine sulphate etc.). In experiments with celite, Nossel et al. (1968) found that the inhibitory substances were adsorbed to the celite particles and thereby prevented the adsorption and consecutive activation of factor XII. Negatively charged substances such as heparin and insulin had no inhibitory effect. This and similar experiments all favour the concept that a negative surface charge is essential for adsorption and activation of factors XII and XI (Haanen et al., 1961; Vroman, 1967).

Isolated factor XII, prepared free from factor XI from normal or factor XI deficient plasma, also becomes activated by adsorption to an appropriate surface. It has been suggested that this activation consists in a reorientation or unfolding of the molecule with the exposure of active enzymic sites (Vroman, 1967, 1968; Margolis, 1963). Activated factor XII differs from its precursor by a different solubility and a different sedimentation pattern on ultracentrifugation (Donaldson and Ratnoff, 1965). It has been suggested that its effect on factor XI involves a limited proteolysis (Schoenmakers et al., 1965).

Several investigators have assumed that factor XII, in agreement with the cascade theory, is activated first, and then follows activation of factor XI, which in turn activates factor IX (Vroman, 1958; Hardisty and Margolis, 1959; Ratnoff et al., 1961; Soulier et al., 1958). However, if factor XII, which has first been adsorbed to a surface, activated and then eluted, is incubated with factor XI in the absence of an activating surface, no clot promoting reaction results. The activation of factor XI requires that both factors simultaneously are adsorbed to an appropriate surface (Hardisty and Margolis, 1959). It has been concluded therefore that factor XIIa and factor XI form a complex on the activating surface during which factor XI is activated. The complex of XIIa and XIa, which also can be eluted from an activating surface, activates factor IX (Vroman, 1958; Waaler, 1959). The coagulant activity of this complex is directly proportional to the amount of factor XI initially present, whereas the rate of reaction depends on the amount of factor XII and the surface (Nossel, 1964). The complex of factor XIIa and factor XIa does not dissociate (Haanen et al., 1967). When factor XIa is inactivated by a normal plasma constituent, an anti-factor XIa-factor, the inactivated factor XIa still remains in the complex with factor XII.

Several investigators have demonstrated that the role of activated factor XII is not limited to the process of blood coagulation but extends to the kinin forming and the fibrinolytic enzyme systems and the complement system

(for references see reviews of EISEN, 1964; RATNOFF, 1966; SHERRY et al.,
1966). Activated factor XII activates plasma prekallikrein to kallikrein, which
subsequently activates plasma kininogen to kinin (MARGOLIS, 1958a, b, 1960;
HABERMAN, 1966; SHIGEHARU et al., 1968). The prekallikrein activator is
probably a fragment of factor XII with a molecular weight of only about
37000 (ØZGE-ANWAR et al., 1972). It has been demonstrated in experimental
animals, that activated Hageman factor could induce increased vascular
permeability (MARGOLIS, 1958b), dilatation of blood vessels (GRAHAM et al.,
1965), and contraction of smooth muscles (MARGOLIS, 1958a). RATNOFF and
MILES (1964) found that purified activated Hageman factor, when added to
diluted plasma, stimulated production of a permeability-producing factor.
These effects are probably mediated by the release of plasma kinins which
produces the characteristics of the inflammatory process (KELLERMEYER and
GRAHAM, 1968). It should be noted that patients with Hageman trait seem
to respond normally to inflammatory processes. Factor XII therefore is not
essential for the responses mentioned (RATNOFF, 1966).

Several investigators have indicated that activated factor XII also activates
plasminogen (NIEWIAROWSKI and PROU-WARTELLE, 1959; IATRIDIS and FER-
GUSON, 1961). IATRIDIS and FERGUSON (1962) and HOLEMANS and ROBERTS
(1964) found reduced fibrinolytic activity in patients with factor XII deficiency
and RATNOFF (1966) has reported defective fibrinolysis in a patient with
factor XI deficiency. Purified factor XII however does not induce fibrinolysis
in vitro. An additional factor, the Hageman cofactor is probably necessary
for this function (OGSTON et al., 1971). It has also been suggested that factor XII is involved in production of acute gouty arthritis (KELLERMEYER
and BRECKENRIDGE, 1965). DONALDSON (1968) has recently found that activated factor XII is capable of transforming the first component of complement
(C_1) to the active form (C_1-esterase) in a purified preparation. He noted that
the factor XII preparation was not free of plasminogen and there is reason
to assume that activated factor XII can activate plasminogen into plasmin
which in turn activates C_1. There are in fact several possibilities of alternative
pathways for the interactions between the various systems mentioned. Activated factor XII may activate the permeability factor which in turn activates
prekallikrein; it may activate plasminogen to plasmin which in turn produces
kinins, and it may activate the complement system. Although the exact
mechanisms and interactions are not clear it is established that activation
of factor XII leads to coagulation, fibrinolysis, activation of the complement
system and the formation of kinins.

The Activation of Factor IX

The formation of the contact or activation product of factors XIIa and XIa
in the presence of an appropriate surface, takes place in the absence of calcium.

From kinetic studies has been concluded that factor XIa is the enzyme which activates factor IX in a time consuming reaction. This activation requires the presence of calcium. The phenomena reported in association with glass contact activation suggest that activation factor XI is not dissociated from the complex with factor XII during its action on factor IX. An alternative explanation has been given by Esnouf (1968) who suggested that factor XI might act as a "coenzyme-like" substance aiding factor XII in a similar way as factor V aids activated factor X in prothrombin conversion. These problems will not be solved until factors XII and XI are isolated in highly purified form.

The appearance of activated factor IX activity as a result of glass contact was first reported by Rapaport et al. (1955). The reaction has since been studied by several groups of investigators. The amount of activated factor IX which is formed is proportional to the initial amount of factor IX, whereas the rate of the reaction is determined by the amount of the activation product (Soulier et al., 1958; Waaler, 1959; Ratnoff and Davie, 1962; Nossel, 1964 and others). Ratnoff and Davie (1962) and Kingdon et al. (1964) found that the activation of factor IX was inhibited by heparin, whereas Pitlick (1968) concluded that heparin inhibits activated factor IX but does not block its formation. It has been reported that activated factor IX differs in electrophoretic mobility from the non-activated, suggesting that a slight proteolysis has taken place (Schiffman et al., 1964). Definite separation and identification of an activated from an inactive factor IX has not been achieved however.

The Activation of Factor X

According to the cascade theory activated factor IX was supposed to activate factor VIII in a simple enzymic reaction, followed by activation of factor X by factor VIIIa. Early studies on the factor X converting activity suggested an interaction of factors IXa and VIII. Calcium was required and phospholipids potentiated the reaction (Biggs et al., 1953a, b; Macfarlane et al., 1964). Studies on the nature of the intrinsic factor X activator have been hampered by its low stability, but recent studies suggest that all the four factors mentioned have to be present simultaneously. They probably form a lipid-calcium-enzyme complex which is the actual activator of factor X (Schiffman et al., 1966; Hougie et al., 1967; Barton, 1967; Hemker et al., 1967; Davie et al., 1969; Hemker et al., 1969). It has further been found that trace amounts of thrombin change factor VIII, possibly by proteolysis, into a more reactive molecule, which accelerates the formation of the factor X activator complex (Rapaport et al., 1963, 1965; Biggs et al., 1965). An increase in activity of 50 times has been recorded (Thompson and Davie, 1971). Larger amounts of thrombin on the other hand are known to produce inactivation of factor VIII by rapid proteolysis. It is not yet known whether a preactivation of factor VIII by traces of thrombin is a prerequisite for the normal formation of the factor X activator. Activated factor VIII has not been isolated, because attempts of separation have resulted in loss of activity.

Factors VIII and IX are adsorbed on the surface of phospholipid micelles in the presence of calcium. BARTON (1967) found that if this complex was rechromatographed in the absence of calcium ions, factor VIII and activated factor IX were recovered unchanged. No activated factor VIII, nor factor X activating activity, could be eluted. It has been suggested therefore that factor VIII only aids factor IXa which is the actual enzyme acting on factor X as the substrate.

The Conversion of Prothrombin to Thrombin

The activation of prothrombin involves a complex mechanism which has created much controversy. The concept that activated factor V is the enzyme which activates prothrombin, as suggested by HARDISTY in 1955 and BRECKEN-RIDGE and RATNOFF (1963, 1964, 1965), and presented in the cascade theory, has not been confirmed.

It has long been suggested that factor V acts together with other conversion factors in prothrombin conversion (OWREN, 1953; JENSEN et al., 1955; GRAY et al., 1956) and SURGENOR et al. (1961) indicated that factor V has a catalytic role in prothrombin activation. It is now well documented that activated factor X is essential for this reaction. The activation with factor Xa alone however, is a very slow process (HOUGIE, 1957; MACFARLANE, 1961; SEEGERS et al., 1962, 1967). Factor Xa, factor V, phospholipids and calcium have all to be present simultaneously for optimal prothrombin conversion. Factor Xa is strongly adsorbed to phospholipid micelles in the presence of Ca^{++} (JOBIN and ESNOUF, 1967; BARTON et al., 1967; PRENTICE et al., 1967; HEMKER et al., 1970), whereas factor V is thought to be bound to phospholipid micelles by its hydrophobic part (KAHN and HEMKER, 1969). Recent evidence has established that the converting moiety, prothrombinase, is a complex of factors Xa and V, Ca^{++} ions and phospholipid. The complex has been visualized as a part of a phospholipid micelle to which one molecule of each of the two clotting factors is adsorbed, one beside the other (HEMKER et al., 1967; JOBIN and ESNOUF, 1967). The form in which factor V functions in the complex has not been definitely established but most investigators believe that factor V becomes activated by thrombin and thereby markedly accelerates the rate of prothrombin conversion (SEEGERS et al., 1947; OWREN, 1947). The activity of factor V, as measured on factor V deficient plasma, increases about 20 times for human factor V and 40 times for bovine factor V after exposure to thrombin (STORMORKEN, 1957). The situation is similar to the activation of factor VIII in the factor X activation complex. Factor V functions like a co-enzyme (although it is not so by definition) and factor Xa is the enzyme (ESNOUF, 1968). It has also been suggested that factor V controls the rate constant of the enzyme substrate complex formation (MAMMEN, 1971) or it may modify the conformation of the substrate (prothrombin) making it more susceptible to the action of the enzyme.

The phospholipid micelles required for the formation of prothrombinase can be furnished by both platelets, erythrocyte stroma and tissue thromboplastin. It contains phosphatidyl ethanolamine and phosphatidyl serine and has a negative surface charge.

During activation of prothrombin to thrombin an intermediate prothrombin fragment has been isolated and termed "prethrombin" (SEEGERS and MARCINIAK, 1965; SEEGERS et al., 1967; TISHKOFF et al., 1968). Pre-thrombin is a smaller molecule than prothrombin. The molecular weight has been calculated to 52000 (MURANO, 1968; TISHKOFF et al., 1968). Pre-thrombin has the two chain configuration of the thrombin molecule. One of the N-terminal amino acids of thrombin, threonine, is exposed. Pre-thrombin could be a laboratory artifact, but on the other hand, it is also possible that additional intermediates might be found (MAMMEN, 1971).

The dissociation of prethrombin to thrombin was extensively studied by MARCINIAK and SEEGERS (1966). Pre-thrombin could be activated by 20% sodium citrate solution, but when activated factor X was added, thrombin formed by a first order reaction with respect to the substrate (SEEGERS and MARCINIAK, 1965).

Thrombin has been purified by several investigators. Highly purified human thrombin was prepared by KORSAN-BENGTSEN and YGGE (1961) by the use of carboxymethyl cellulose. MAGNUSSON (1965) prepared highly purified bovine thrombin by means of ethanol fractionation and chromatography on Amberlite IRC-50. Several other methods for purification of thrombin have been described (for references see MURANO, 1971). Even highly purified thrombin preparations contain multiple components (ROZENBERG and WAUGH, 1970). THOMPSON and DAVIE (1971) by affinity chromatography, prepared bovine thrombin free of contaminant inhibitors, as well as other similar esterases including activated factor X, with a specific activity of 2500 N.I.H. units per mg protein.

The molecular weight of human thrombin has been estimated to 35000 (LANCHANTIN et al., 1965) and 26000–32000 (MAGNUSSON, 1965f). Thrombin has esterase and clotting activities which are both associated with a single protein component (LANCHANTIN et al., 1965). The kinetics of the hydrolysis of several synthetic substrates by human thrombin have been reported by KEZDY et al. (1965).

Bovine thrombin has two N-terminal amino acids, isoleucine and threonine, which indicate that the thrombin structure consists of at least two polypeptide chains (MAGNUSSON, 1965 d). The primary structure of bovine thrombin has been analyzed by MAGNUSSON (1969). He has suggested that prothrombin consists of a single polypeptide chain which is changed to a double chain in thrombin.

Thrombin causes a limited proteolysis of the fibrinogen molecule. It has a predilection for arginyl and lysyl bonds which makes the specificity of thrombin similar to that of trypsin, although it is much more selective. In fibrinogen it splits only four of more than one hundred bonds which are potentially susceptible to trypsin.

The Fibrinogen-Fibrin Conversion

BAILY et al. in 1951 observed that the interaction of purified thrombin and fibrinogen consistently produced new terminal residues of glycine confirming the old suggestion that thrombin is a specific proteolytic enzyme. In the same year LORAND discovered that the fibrinopeptides were split off from fibrinogen by the action of thrombin. It was soon confirmed that thrombin during the initial reaction releases fibrinopeptides A and B which were identified in clot supernatant solutions (BETTELHEIM and BAILY, 1952; LORAND, 1952).

It has already been mentioned that the fibrinogen molecule is a dimeric unit, each unit consisting of three peptide chains α (A), β (B) and γ (CLEGG and BAILEY, 1962; HENSCHEN, 1964a). At 37°C, the dimer structure is dissociated into two subunits of a molecular weight of 180000 (CAPET-ANTONINI and GUINAND, 1967). The N-terminal ends of the three chains are linked together in a firm disulfide knot (BLOMBÄCK et al., 1968). Thrombin splits off four fibrinopeptide fragments by breaking four arginyl-glycine bonds (BLOMBÄCK, 1967). The fibrinopeptides are split off from the N-terminal ends of the α (A) and β (B) chain. They are denoted peptide A and peptide B. The release of peptide A occurs more rapidly than that of peptide B (BLOMBÄCK and BLOMBÄCK, 1969).

When the fibrinopeptides A and B are split from fibrinogen the resulting fibrin monomers start aggregating. Certain structures in the disulfide knot are of importance for the polymerization (BLOMBÄCK et al., 1969). The aggregation of the fibrin monomers is controlled by electrostatic forces and not by covalent bonds. Both end to end and side by side associations are involved in the soluble gel formation. Otherwise, the incipient polymerization of the fibrin monomers is not wellunderstood. Agents which are known to interfere with hydrogen bond formation, such as urea, can disrupt the gel at this stage of reaction.

The soluble fibrin gel is reinforced by the introduction of covalent bonds which crosslink the constituent molecules. This stabilization, which is necessary for fulfilling the physiological function of fibrin, is catalyzed by factor XIII. In order to exhibit this function, factor XIII has first to be activated by thrombin, as demonstrated by Buluk et al. (1961). The activation is caused by a limited proteolytic effect which uncovers new sulfhydryl groups (Buluk et al., 1966). Whether small amounts of peptide material are released is not known. Both activation of factor XIII and the action of factor XIIIa proceed at high velocity. Calcium ions are required for the activation of factor XIII (Lorand and Konishi, 1964). The actual nature of the crosslink in fibrin was demonstrated by three research groups at about the same time (Matacic et al., 1968; Pisano et al., 1968; Lorand et al., 1968). It takes place by the formation of γ-glutamyl lysine bonds between fibrin units, during which ammonia is released.

The Extrinsic Pathway of Coagulation

The formation of extrinsic prothrombinase requires tissue thromboplastin, factors VII, X, V and calcium ions. Several studies using sera from patients with congenital deficiencies and purified preparations of factors VII and X demonstrated that the interaction of tissue thromboplastin and factors VII and X, in the presence of calcium, resulted in an "extrinsic reaction product" which activated prothrombin to thrombin in the presence of factor V, phospholipid and calcium (Hjort, 1957; Hougie, 1959; Straub and Duckert, 1961; Nemerson and Spaet, 1964; Nemerson, 1966). Tissue thromboplastin and factor VII in the presence of calcium formed a complex which sedimented with the thromboplastin particle (Hjort, 1957; Nemerson, 1966). The tissue thromboplastin-factor VII-calcium complex could not be dissociated by removal of calcium and neither factor VII nor factor X activator could be recovered in the supernatant after high speed centrifugation. Recent studies suggest that factor VII is activated and combines with the phospholipid part of the tissue thromboplastin (Williams and Norris, 1966; Østerud, 1972). After activation of factor X the pathway is the same as for the intrinsic clotting systems.

The Activation of Coagulation in Vivo

Tissue thromboplastin is an intracellular substance which is released in association with tissue damage and initiates the extrinsic clotting system. The normal bleeding time in haemophilia, in which the intrinsic clotting system is out of function, indicates that the extrinsic system together with platelets can provide for provisional haemostasis. The normal bleeding time in factor VII deficiency illustrates on the other hand that the intrinsic clotting system is also triggered by tissue damage but the mechanism is obscure. It has been found that both collagen and elastin activate factor XII (Niewiarowski et al.,

1964, 1965; WILNER et al., 1968). Collagen also induces platelet adhesion and the release reaction, providing ADP for aggregation and phospholipid (platelet factor 3) for coagulation (HOLMSEN et al., 1969). By exposure of connective tissue to the blood stream by intimal damage therefore, seems favourable for both formation of haemostatic plugs in wounds and for thrombus formation. However, in arterial thrombosis it has been found that polymerized fibrin is absent at just the sites where coagulation would be anticipated, i.e. at the contact area between the platelet thrombus and the vessel wall. Fibrin appears only at the surface of the platelet thrombus facing the blood stream (KJÆR-HEIM and HOVIG, 1962). In venous thrombosis there is no evidence of endo-thelial damage which possibly could release thromboplastin from the sub-endothelial tissue or expose collagen fibres. It is unlikely therefore that coagula-tion is activated by substances from the vessel wall in this case.

In 1955 HJORT et al. demonstrated that factor V from plasma is strongly adsorbed on the platelet surface, producing the activity which was previously known as platelet factor 1. It has since been demonstrated that all the plasma clotting factors are represented on the platelet surface. The clinical signifi-cance of the platelet surface clotting system for haemostasis was demon-strated several years ago when it was found that transfusions of normal platelets temporarily restored the prolonged secondary bleeding time in haemo-philia A and B and factor V deficiency and also the prolonged primary bleeding time in patients who had been overtreated with anticoagulants (BORCH-GREVINK and OWREN, 1961). Platelet transfusions had no effect on the plasma clotting systems and it was concluded therefore that clotting factors on the surface of the transfused platelets provided thrombin for haemostasis.

The trigger mechanism for platelet surface coagulation has remained un-known. WALSH (1972) recently demonstrated that something in the platelets can initiate intrinsic coagulation by enhancing the formation of the activation product from factors XII and XI. The effect of this "contact product forming activity" (CPFA) was completely dependent upon the presence of factor XII and partially dependent upon the presence of factor XI. It seems possible therefore that whenever the platelet release reaction is initiated, CPFA is released and activates intrinsic coagulation on the platelet surface.

There are still many questions which are not answered by the discovery of CPFA, because the effect of this factor requires the presence of factor XII. Since factor XII deficiency exhibits no bleeding tendency there are obviously mechanisms in the body for direct activation of factor XI.

It is quite likely that this trigger is liberated from the platelets during the release reaction, but an "intrinsic trigger", with an activating effect directly on factor XI has not been identified. Platelet phospholipids (platelet factor 3) which is exposed on the platelet surface during the release reaction, activates factor XII (LÜSCHER, 1968), which like other clotting factors is adsorbed on the platelet surface (Sharp, 1958; Waaler, 1959). But it has not been confirmed that phospholipids can activate factor XI directly as has been suggested (Rapaport, 1959) and injection of even large amounts of disrupted platelets gives negligible stimulation of coagulation (Evensen and Jeremic, 1970).

Extensive research has been devoted to finding the trigger mechanism of in vivo coagulation in disseminated intravascular coagulation. This is a well known complication in bacterial shock, meningeoccal septicemia, haemolytic syndromes, severe trauma, malignant disease etc. and is characterized by microthrombosis and/or haemorrhage. The condition has also been termed consumption coagulopathy (Lasch et al., 1967; Roka, 1971). Intravascular coagulation can be produced by tissue thromboplastin (Izak and Galewsky, 1966; Schneider, 1969), by endotoxin (Des Prez et al., 1961; Lipinsky et al., 1968; McKay, 1965), by agents which activate factor XII (Botti and Ratnoff, 1964; Rodriguez-Erdmann, 1964; Müller-Berghaus and Lasch, 1970), by infusion of thrombin (Margaretten et al., 1964), by certain exogenous proteolytic enzymes (McKay, 1965), by proteolytic enzymes from leucocytes (Horn and Collins, 1968; Huth et al., 1968) and from endothelial cells (Müller-Berghaus and Lasch, 1963; McGrath and Stewart, 1969). Trypsin, certain snake venoms and other proteolytic enzymes might also trigger the coagulation system in vivo. Liquoid, which is a synthetic acid polymer with an antithrombin effect, also triggers intravascular coagulation and causes generalized Schwartzman reaction in rabbits, even when they are made thrombocytopenic (Evensen et al., 1967; Evensen and Jeremic, 1968). The effect might be mediated through an activation of factor XII (Müller-Berghaus and Lasch, 1971). Whether intravascular coagulation results from these various agents depends largely on the activity of the fibrinolytic system, the state of the adrenergic system and the efficiency of the reticulo-endothelial system. Inhibition of fibrinolysis (by epsilon-aminocaproic acid), inhibition of the reticulo-endothelial system and stimulation of the adrenergic system by norepinephrine favour intravascular coagulation. It has been suggested that certain agents activate the intrinsic or extrinsic clotting system directly, others release trigger substances from platelets, endothelial cells, leucocytes or the tissues.

Interactions between the Extrinsic and the Intrinsic Pathways of Coagulation

The activity of factor VII, as measured in thromboplastin time systems, increases by contact activation of plasma by glass. Waaler (1959) and Soulier and Prou-Wartelle (1960) found that factor XII was compulsory for this process. Altman and Hemker (1967), by using the thrombotest method for analyzing the degree of contact activation, confirmed that activated factor XII can convert factor VII into its active form. Factor XI markedly enhanced the reaction without being indispensable. The significance of this activation in vivo is unknown but it has been speculated that it favours the production of small amounts of thrombin by the extrinsic system which can activate factors VIII and V and thereby enhance or short circuit the intrinsic pathway (RAPAPORT et al., 1965).

Tissue extract, in the absence of contact with glass, can promote thrombin generation by the intrinsic clotting system. Factors XII and XI do not seem necessary for this effect, but factor VII is essential. The complex formed by tissue thromboplastin, factor VII and calcium, can substitute for the activation product of factors XII and XI in the activation of factor IX (JOSSO and PROU-WARTELLE, 1965). It has been suggested that this mechanism of activation of factor IX, which bypasses factor XII, under certain circumstances may be of importance in factor XII deficiency patients. The bleeding tendency in factor XI deficiency however, shows that this type of activation of the intrinsic clotting system cannot compensate for other normal in vivo triggering mechanisms.

Coagulation Inhibitors

There seems to be plasma inhibitors for most of the activated products occurring during coagulation. These inhibitors have not been thoroughly studied except the antithrombins, although it is now realized that both natural and pathological inhibitors have important clinical implications. Even a moderate congenital deficiency of antithrombin 3 is associated with a definite thrombotic tendency (EGEBERG, 1965) and pathological inhibitors can cause a bleeding tendency (ROBINSON et al., 1967).

HOWELL (1916–1917) suggested that thrombin was inactivated by its combination with a specific protein (antithrombin) forming a complex which he termed metathrombin. By the use of purified preparations ABILDGAARD (1969) varied the concentrations of enzyme and inhibitor, and based on molecular weights, he observed the formation of a binary complex.

Six natural antithrombins have been described. Antithrombin 1 refers to the adsorption of thrombin to the fibrin gel (SEEGERS et al., 1945). Antithrombin 2 is the plasma cofactor of heparin (BRINKHOUS et al., 1939). Antithrombin 3, often referred to as the progressive antithrombin, is the capacity of plasma to inactivate thrombin. Antithrombin 4 has been described as the ability of plasma to inactivate freshly formed thrombin, but not "stabilized" thrombin. Antithrombin 5 is an activity associated with the γ-globulins in certain diseases like myelomatosis and rheumatoid arthritis (LOELIGER and HERS, 1957; VINAZZER and REINHARDT, 1968; SCHIMPF, 1971). Antithrombin 6 is the inhibitory effect of fibrinogen-fibrin split products on the thrombin-fibrinogen reaction (Niewiarowski and Kowalski, 1957, 1958; Kowalski, 1959).

Antithrombin 3 has been concentrated from plasma and serum and some of its physico-chemical characteristics have been reported (Seegers et al., 1964a). Recent investigations have demonstrated that antithrombin 2 (heparin co-factor) is one and the same substance as antithrombin 3 (Lyttleton, 1954; Burstein, 1955; Monkhouse et al., 1955; Abildgaard, 1968). Abildgaard (1967) and Ganrot and Niléhn (1969) have shown that the activity of anti-thrombin 3 is due to two substances, an α_2-globulin and an α_2-macroglobulin, the first representing the antithrombin 2 and 3 activity.

Investigations by Dombrose et al. (1971) showed that the binding of anti-thrombin to thrombin followed a second order reaction that could be dis-tinguished into an essentially irreversible phase and a slower, potentially reversible phase.

Extensive work by Seegers et al., with highly purified preparations have demonstrated that activated factor X is also inhibited by purified anti-thrombin in much the same way as thrombin (Seegers et al., 1964a). These findings have been confirmed by Egeberg (1965a, b), Biggs et al. (1970) and Yin and Wessler (1970). The inhibitory effect on activated factor X has often been referred to as anti-plasma-thromboplastin.

The effect of fibrinogen split products (antithrombin 6) is non-specific and interact mainly with fibrin polymerization (Niewiarowski and Kowalski, 1957; Triantaphyllopoulos, 1969; Verstraete et al., 1971).

Normal plasma and serum also contain an inhibitor against activated factor XI (Margolis, 1957; Ratnoff et al., 1961; Nossel and Niemetz, 1965). The inhibitor was purified from plasma and it was demonstrated that it specifically blocked factor XI activity (Niemetz and Nossel, 1967). Activated factors VIII and V are both rapidly inactivated by thrombin. Human serum therefore, is devoid of both factor VIII and factor V. Johnson and Seegers (1954) and Mammen et al. (1960b) have suggested that factor VIII combines with a lipid during coagulation and is thereby inactivated. The formation of this complex was assumed to be mediated by thrombin. Factor VIII was recovered from serum after ether extraction of the lipid (Mammen, 1965).

The physiological significance of the natural inhibitors is obvious. Coagulation in vivo has to be restricted to serve haemostatic and other local purposes. This function requires a balanced system of procoagulant and anticoagulant activities, and the latter might be just as important as the clot promoting factors.

In addition to the plasmatic inhibitors mentioned it is important to realise that the organism also has an effective protection in the reticulo-endothelial system which quickly removes coagulation products which enter the general circulation. The importance of this system has been demonstrated by experimental studies on disseminated intravascular coagulation (EVENSEN and HJORT, 1970; see monographs by McKAY, 1965; HARDAWAY, 1966).

Pathological circulating anticoagulants or inhibitors, which may occur in many disorders, are also of great clinical interest. They can produce bleeding disorders or worsen an existing haemorrhagic disease which might become refractory to substitution therapie, such as in haemophilia. The latter Type of inhibitors usually develop after treatment with plasma or plasma fractions, and are isoantibodies. This type of inhibitors occurs predominantly in patients with severe congenital deficiencies of factors VIII or IX. They are predominantly gamma-globulins of the IgG class (for references see LECHNER, 1969). Inhibitors of a different nature may develop in collagen diseases (disseminated lupus erythematosis, rheumatoid arthritis), dermatological diseases (pemphigus vulgaris, dermatitis herpetiformis), liver diseases, diseases of the lymphatic system and as a result of a drug allergy. In amyloidosis, an inhibitor against factor X appears frequently.

Of particular interest are the coagulation inhibitors which occur during anticoagulant therapy and in K-avitaminosis and which has been termed PIVKA (protein induced by vitamin K absence or antagonists) (HEMKER, 1964; HEMKER et al., 1968; HEMKER and MULLER, 1968). These inhibitors act as competitive inhibitors of prothrombin conversion. They have as already discussed, been identified as prestages of prothrombin and factors VII, IX and X. This discovery has obvious consequences for the management and control of anticoagulant therapy. It has been found that a control method, such as Thrombotest (OWREN, 1959), which is sensitive to these inhibitors, records about half the percentage values of those recorded by insensitive methods, such as the one-stage prothrombin time test (OWREN, 1962, 1963, 1971).

It has already been mentioned that molecular abnormalities of the clotting factors, which can be detected immunologically, are present in a large number of so-called deficiency states. It has been observed in haemophilia A and B and factor X deficiency. These defects seem to vary considerably, suggesting that each patient to a certain extent manufactures his own type of molecular alteration of the factor in question (BRECKENRIDGE and HOYER, 1971; LARRIEU and MEYER, 1971; DENSON, 1971). Several congenital fibrinogen anomalies have also been reported (MÉNACHÉ, 1964; FISCHER et al., 1968; BLOMBÄCK et al., 1968; MÉNACHÉ, 1970). Molecular abnormalities presumably exist also for the other deficiency states. Some of these abnormal molecules act as inhibitors such as illustrated by haemophilia Bm.

The inhibitory effect of fibrinogen-fibrin split products is also clinically important. Such inhibitory effect might be observed even in mild cases of disseminated intravascular coagulation and fibrinolysis as seen postoperatively,

by injury, burns, in malignant diseases, liver diseases, postpartum etc. Muirhead and Triantaphyllopoulos (1971) found that thrombin also lysed fibrin
to similar breakdown products as those formed during the lysis of fibrin
and fibrinogen by plasmin. The split products of fibrin and fibrinogen exhibit
antithrombotic effect and inhibit the consumption of prothrombin (Triantaphyllopoulos et al., 1969).

References

Aas, K.: Prokonvertin og konvertin. Undersøkelser over blodets koagulasjon med spesielt
    henblikk på prokonvertin og konvertin. Thesis. Akad. trykningssental Oslo 1952,
    p. 351.
Abildgaard, U., Purification of two progressive antithrombins of human plasma. Scand.
    J. clin. Lab. Invest. 19, 190 (1967).
Abildgaard, U.: Highly purified antithrombin III with heparin cofactor activity pre-
    pared by disc electrophoresis. Scand. J. clin. Lab. Invest. 21, 89 (1968).
Abildgaard, U.: Binding of thrombin to antithrombin III. Scand. J. clin. Lab. Invest.
    24, 23 (1969).
Addis, T.: The pathogenesis of hereditary haemophilia. J. Path. Bact. 15, 427 (1911).
Aggeler, P. M., White, S. G., Glendening, M. B., Page, E. W., Leake, T. B., Bates,
    G.: Plasma thromboplastin component (PTC) deficiency: a new disease resembling
    haemophilia. Proc. Soc. exp. Biol. (N.Y.) 79, 692 (1952).
Alexander, B.: Some biochemical, physicochemical and immunochemical studies of
    prothrombin and proconvertin (factor VII): their biopathologic significance. In:
    Blood clotting factors (E. Deutsch, ed.), p. 37. London: Pergamon Press 1958.
Alexander, B., Colman, R.: Evidence for a new serum thromboplastic factor. Throm-
    bos. Diathes. haemorrh. (Stuttg.) 4 (Suppl.), 66 (1959).
Alexander, B., De Vries, A., Goldstein, R., Landwehr, G.: Prothrombin conversion
    accelerator in serum. Science 109, 544 (1949).
Alexander, B., Goldstein, R., Landwehr, G., Cook, C. D.: Congenital SPCA defi-
    ciency: a hitherto unrecognized coagulation defect with hemorrhage rectified by
    serum and serum fractions. J. clin. Invest. 30, 596 (1951).
Altman, R., Hemker, H. C.: Contact activation in the extrinsic blood clotting system.
    Thrombos. Diathes. haemorrh. (Stuttg.) 18, 525 (1967).
Aoki, N., Harmison, C. R., Seegers, W. H.: Properties of bovine Ac-globulin concen-
    trates and methods of preparation. Canad. J. Biochem. 41, 2409 (1963).
Arthus, M., Pages, C.: Nouvelle théorie chimique de la coagulation du sang. Arch.
    Physiol. norm. path. 2, 739 (1890).
Bailey, K., Bettelheim, F. R., Lorand, L., Middlebrook, W. R.: Action of thrombin
    in the clotting of fibrinogen. Nature (Lond.) 167, 233 (1951).
Bang, N. U.: A molecular structural model of fibrin based on electron microscopy of
    fibrin polymerization. In: Fibrinogen and fibrin. Turnover of clotting factors. Throm-
    bos. Diathes. haemorrh. (Stuttg.) 13 (Suppl.), 73 (1964).
Bangham, A. D.: A correlation between surface charge and coagulant action of phospho-
    lipids. Nature (Lond.) 192, 1197 (1961).
Barnhart, M. J., Cress, D. C., Noonan, S. M., Walsh, R. T.: Influence of fibrinolytic
    products on hepatic release and synthesis of fibrinogen. Thrombos. Diathes. haemorrh.
    (Stuttg.) 39 (Suppl.), 143 (1970).
Barrow, E. M., Amos, S., Graham, J. B.: Certain biochemical properties of human AHF
    (Factor 8) separated from fibrinogen with manganous chloride and thrombin. J. Lab.
    clin. Med. 68, 803 (1966).
Barrow, E. M., Graham, J. B.: Kidney antihemophilic factor. Partial purification and
    some properties. Biochemistry 7, 3917 (1968).
Barton, P. G.: Sequence theories of blood coagulation reevaluated with reference to
    lipid-protein interaction. Nature (Lond.) 215, 1508 (1967).
Barton, P. G., Hanahan, D. J.: The preparation and properties of a stable factor V
    from bovine plasma. Biochim. biophys. Acta (Amst.) 133, 506 (1967).

BARTON, P. G., JACKSON, C. M., HANAHAN, D. J.: Relation between factor V and activated factor X in the generation of prothrombinase. Nature (Lond.) **214**, 923 (1967).

BERGSAGEL, D. E., NOCKOLDS, E. R.: The activation of proaccelerin. Brit. J. Haemat. **11**, 395 (1965).

BERGSTRÖM, D., WALLÉN, P.: Removal of contaminating plasminogen from purified bovine fibrinogen. Arkiv Kemi **17**, 503 (1961).

BETTELHEIM, F. R., BAILEY, K.: The products of the action of thrombin on fibrinogen. Biochim. biophys. Acta (Amst.) **9**, 578 (1952).

BIGGS, R.: Report on the standardization of the one-stage prothrombin time for the control of anticoagulant therapy. Thrombos. Diathes. haemorrh. (Stuttg.), Suppl. **17**, 303 (1965).

BIGGS, R.: Proposal for the use of a preliminary trial reference preparation of thromboplastin for standardizing the one-stage prothrombin time for the control of anticoagulant therapy. Thrombos. Diathes. haemorrh. (Stuttg.), Suppl. **39**, 363 (1970).

BIGGS, R., BIDWELL, E.: An attempt to identify a single phospholipid active in blood coagulation. Brit. J. Heamat. **3**, 387 (1957).

BIGGS, R., DENSON, K. W. E.: Second report on the standardization of the one-stage prothrombin time for the control of anticoagulant therapy. Thrombos. Diathes. haemorrh. (Stuttg.), Suppl. **20**, 345 (1966).

BIGGS, R., DENSON, K. W. E.: Standardization of the one-stage prothrombin-time of anticoagulant therapy. Brit. med. J. **1967 I**, 84.

BIGGS, R., DENSON, K. W. E., AKMAN, N., BORRETT, R., HADDEN, M.: Antithrombin III, antifactor Xa and heparin. Brit. J. Haemat. **19**, 283 (1970).

BIGGS, R., DOUGLAS, A. S., MACFARLANE, R. G.: The formation of thromboplastin in human blood. J. Physiol. (Lond.) **119**, 89 (1953a).

BIGGS, R., DOUGLAS, A. S., MACFARLANE, R. G.: The initial stages of blood coagulation. J. Physiol. (Lond.) **122**, 538 (1953b).

BIGGS, R., DOUGLAS, A. S., MACFARLANE, R. G., DACIE, J. V., PITNEY, W. R., MERSKEY, C., O'BRIEN, J. R.: Christmas disease: a condition previously mistaken for haemophilia. Brit. med. J. **1952 II**, 1378.

BIGGS, R., MACFARLANE, R. G., eds.: Human blood coagulation and its disorders, p. 474. Oxford: Blackwell 1962.

BIGGS, R., MACFARLANE, R. G., DENSON, K. W. E., ASH, B. J.: The interaction of factors VIII and IX. Brit. J. Haemat. **11**, 276 (1965).

BIGGS, R., MATTHEWS, J. M.: The treatment of haemorrhage in von Willebrand's disease and the blood level of factor VIII (AHG). Brit. J. Haemat. **9**, 203 (1963).

BIGGS, R., SHARP, A. A., MARGOLIS, J., HARDISTY, R. M., STEWART, J., DAVIDSON, W. M.: Defects in the early stages of blood coagulation: a report of four cases. Brit. J. Haemat. **4**, 177 (1958).

BLAINVILLE, DE: 1834. Quoted by NYGAARD, K. K.: Hemorrhagic diseases. St. Louis 1941.

BLOMBÄCK, B.: Fibrinogen to fibrin transformation. In: Blood clotting enzymology, ed. W. H. SEEGERS, p. 143. New York: Acad. Press Inc. 1967.

BLOMBÄCK, B., BIRGER, M., HESSEL, B., IWANAGA, S., REUTERBY, J., BLOMBÄCK, M.: Primary structure of human fibrinogen and fibrin. I. Cleavage of fibrinogen with cyanogen bromide. Isolation and characterization of NH_2-terminal fragments of the α ("A") chain. J. biol. Chem. **247**, 1496 (1972).

BLOMBÄCK, B., BLOMBÄCK, M.: Purification of human and bovine fibrinogen. Arkiv Kemi **10**, 415 (1956).

BLOMBÄCK, B., BLOMBÄCK, M.: A method for the assay of factor V. Scand. J. clin. Lab. Med. **15**, 639 (1963).

BLOMBÄCK, B., BLOMBÄCK, M.: The formation of the fibrin clot from fibrinogen. In: Human blood clotting, ed. HEMKER et al., p. 7. Leiden: University Press 1969.

BLOMBÄCK, B., BLOMBÄCK, M., EDMAN, P. et al.: Human fibrinopeptides. Isolation, characterization and structure. Biochim. biophys. Acta (Amst.) **115**, 371 (1966).

BLOMBÄCK, B., BLOMBÄCK, M., HENSCHEN, A., HESSEL, B., IWANAGA, S., WOODS, R.: N-terminal disulphide knot of human fibrinogen. Nature (Lond.) **218**, 130 (1968).

BLOMBÄCK, B., BLOMBÄCK, M., MAMMEN, E. F., PRASAD, A. S.: Fibrinogen Detroit— a molecular defect in the N-terminal disulphide knot of human fibrinogen? Nature (Lond.) **218**, 134 (1968).

Blombäck, B., Blombäck, M., Nilsson, I. M.: Note on the purification of human antihemophilic globulin. Acta chem. scand. 12, 1878 (1958).

Blombäck, B., Yamashina, J.: On the N-terminal amino acids in fibrinogen and fibrin. Arkiv Kemi 12, 299 (1958).

Blombäck, M.: Purification of antihaemophilic globulin. Arkiv Kemi 12, 387 (1958).

Blombäck, M., Blombäck, B.: Hemophilia and other haemorrhagic states, p. 242. Chapel Hill Univ. N. Carolina Press 1970.

Borchgrevink, C. F., Owren, P. A.: The hemostatic effect of normal platelets in hemophilia and factor V deficiency. Acta med. scand. 170, 375 (1961).

Bordet, J., Delange, L.: Sur la nature du cytozyme. Recherches sur la coagulation du sang. Ann. Inst. Pasteur 27, 341 (1913).

Bordet, J., Delange, L.: Analyse et synthese du processus de la coagulation. Ann. Soc. roy. Sci. méd. nat. Brux. 72, 87 (1914).

Bordet, J., Gengou, O.: Recherches sur la coagulation du sang: contribution de l'étude du plasma fluore. Ann. Inst. Pasteur 17, 822 (1903).

Bordet, J., Gengou, O.: Recherches sur la coagulation du sang et les serum anti-coagulants. Ann. Inst. Pasteur 18, 98 (1904).

Botti, R. E., Ratnoff, O. D.: Studies on the pathogenesis of thrombosis: An experimental "hypercoagulable" state induced by the intravenous injection of ellagic acid. J. Lab. clin. Med. 64, 385 (1964).

Bouma, B. N., Wiegerinck, Y., Sixma, J. J., Mourik, J. A., van, Mochtar, I. A.: Immunological characterization of purified antihaemophilic factor A (factor VIII) which corrects abnormal platelet retention in von Willebrand's disease. Nature (Lond.) New Biol. 236, 104 (1972).

Breckenridge, R. T., Hoyer, L. W.: The hererogeneity of hemophilia. Thrombos. Diathes. haemorrh. (Stuttg.), Suppl. 43, 5 (1971).

Breckenridge, R. T., Ratnoff, O. D.: Studies on the site of action of circulating anticoagulant in disseminated lupus erythematosus. Evidence that this anticoagulant inhibits the reaction between activated Stuart factor (factor X) and proaccelerin (factor V). Amer. J. Med. 35, 813 (1963).

Breckenridge, R. T., Ratnoff, O. D.: Studies on the anticoagulant action of soybean trypsin inhibitor. Clin. Res. 12, 221 (1964).

Breckenridge, R. T., Ratnoff, O. D.: The activation of human proaccelerin (Factor V). Thrombos. Diathes. haemorrh. (Stuttg.) 17, 217 (1965).

Brinkhous, K. M.: A study of the clotting time in hemophilia: the delayed formation of thrombin. Amer. J. med. Sci. 198, 509 (1939).

Brinkhous, K. M., Shanbrom, E., Roberts, H. R., Webster, W. P., Fekete, L., Wagner, R. H.: A new high potency glycine precipitated antihemophilic factor (AHF) concentrate. Treatment of classical hemophilia and hemophilia with inhibitors. J. Amer. med. Ass. 205, 613 (1968).

Brinkhous, K. M., Smith, H. P., Warner, E. D., Seegers, W. H.: The inhibition of blood clotting: an unidentified substance which acts in conjunction with heparin to prevent the conversion of prothrombin into thrombin. Amer. J. Physiol. 125, 683 (1939).

Brown, M., Rothstein, F.: Fibrinogen from human plasma: Preparation by precipitation with heavy-metal coordination complex. Science 155, 1017 (1967).

Buchanan, A.: On the coagulation of the blood and other fibrinoferous liquids. J. Physiol. (Lond.) 2, 158 (1879). (Reprinted from London Med. Gazette, 1845, 1, 617.)

Bulloch, W., Fildes, P.: Treasury of human inheritance. Parts V and VI. Section XIV a. Haemophilia. London: Kulau and Co., Limited, 1911.

Buluk, K., Januszko, T., Olbromski, J.: Conversion of fibrin to desmofibrin. Nature (Lond.) 191, 1093 (1961).

Buluk, K., Olbromski, J., Januszko, T., Zuch, A.: Desmofibrin formation and the activity of the fibrin stabilizing factor (FSF) during the cleavage of its SH groups by thrombin. Thrombos. Diathes. haemorrh. (Stuttg.) 16, 51 (1966).

Burstein, M.: Sur l'activation de la thrombin par le plasma héparine. Arch. int. Pharmacodyn. 101, 285 (1955).

Caldwell, M. J., Kaulla, K. N., von, Seegers, W. H.: Procoagulant material from human urine in prothrombin activation. Thrombos. Diathes. haemorrh. (Stuttg.) 9, 53 (1963).

CAPET-ANTONINI, F., GUINAND, S.: Effet de la température sur la structure du fibrinogene. C. R. Acad. Sci. (Paris) 265, 2093 (1967).

CASPARY, E. A., KEKWICK, R. A.: Some physicochemical properties of human fibrinogen. Biochem. J. 67, 41 (1957).

CHARGAFF, E.: The coagulation of blood. Advanc. Enzymol. 5, 31 (1945).

CHARGAFF, E., BENDICH, A., COHEN, S. S.: The thromboplastic protein: structure, properties, disintegration. J. biol. Chem. 156, 161 (1944).

CLEGG, J. B., BAILEY, K.: The separation and isolation of the peptide chains of fibrin. Biochim. biophys. Acta (Amst.) 63, 525 (1962).

COHEN, C., SLAYTER, H., GOLDSTEIN, L., KUCERA, J., HALL, C.: Polymorphism in fibrinogen aggregates. J. molec. Biol. 22, 385 (1966).

COHN, E. J., STRONG, L. E., HUGHES, W. L., JR., MULFORD, D. J., ASHWORTH, J. N., MELIN, M., TAYLOR, H. L.: Preparation and properties of serum and plasma proteins. IV. A system for the separation into fractions of the protein and lipoprotein components of biological tissues and fluids. J. Amer. chem. Soc. 68, 459 (1946).

CONNOR, W. E., WARNER, E. D., CARTER, J. R.: A labile serum factor clotting defect: its demonstration by the thromboplastin generation test and its clinical significance. J. clin. Invest. 40, 13 (1961).

COX, F. M., LANCHANTIN, G. F., WARE, A. G.: Chromatographic purification of human serum accelerator globulin. J. clin. Invest. 35, 106 (1956).

DAVIE, E. W., HOUGIE, C., LUNDBLAD, R. L.: The mechanism of blood coagulation. In: Recent advances in blood coagulation, ed. POLLER, J. V. A., p. 13. London: Churchill 1969.

DENIS, P. S.: Memoire sur le sang. Paris 1859.

DENSON, K. W. E.: Structurally defective forms of factor VIII, factor IX and factor X. Thrombos. Diathes. haemorrh. (Stuttg.), Suppl. 43, 19 (1971).

DENSON, K. W. E., BIGGS, R., HADDON, M. E., BORRET, E., COBB, K.: Two types of haemophilia (A+ and A-). A study of 48 cases. Brit. J. Haemat. 17, 163 (1969).

DENSON, K. W. E., BIGGS, R., MANNUCCI, P. M.: An investigation of three patients with Christmas disease due to an abnormal type of factor IX. J. clin. Path. 21, 160 (1968).

DENSON, K. W. E., LURIE, E. A., DE CATALDO, F. D. E., MANNUCCI, P. M.: Factor X defect. Recognition of abnormal forms of factor X. Brit. J. Haemat. 18, 317 (1970).

DESAI, R. G.: Hemophilia treated with spleen cells. J. Amer. med. Ass. 207, 1269 (1969).

DES PREZ, R. M., HOROWITZ, H. K., HOOK, E. W.: Effects of bacterial endotoxin on rabbit platelets. I. Platelet aggregation and release of platelet factors in vitro. J. exp. Med. 114, 857 (1961).

DEUTSCH, E., SCHADEN, W.: Zur Reinigung und Charakterisierung des VII. Blutgerinnungsfaktors. Biochem. Z. 324, 266 (1953).

DEUTSCH, E., IRSIGLER, K., LOMOSCHITZ, H.: Studien über Gewebethromboplastin. I. Reinigung, chemische Charakterisierung und Trennung in einen Eiweiß- und Lipoidanteil. Thrombos. Diathes. haemorrh. (Stuttg.) 12, 12 (1964).

DEUTSCH, E., LECHNER, K., SCHMER, G.: On the purification of factor VII from bovine plasma. Thrombos. Diathes. haemorrh. (Stuttg.), Suppl. 20, 275 (1966).

DE VRIES, A., ALEXANDER, B., GOLDSTEIN, R.: A factor in serum which accelerates the conversion of prothrombin to thrombin: I. Its determination and some physiologic and biochemical properties. Blood 4, 247 (1949).

DOMBROSE, F. A., SEEGERS, W. H., SEDENSKY, J. A.: Antithrombin. Inhibition of thrombin and autoprothrombin C (F-Xa) as a mutual depletion system. Thrombos. Diathes. haemorrh. (Stuttg.) 26, 103 (1971).

DONALDSON, V. H.: Mechanisms of activation of C'1 esterase in hereditary angioneurotic edema plasma in vitro. J. exp. Med. 127, 411 (1968).

DONALDSON, V. H., RATNOFF, O. D.: Hageman factor alterations on physical properties during activation. Science 150, 754 (1965).

DUCKERT, F., FLÜCKIGER, P., KOLLER, F.: Le role du facteur X dans la formation de la thromboplastine sanguine. Rev. Hémat. 9, 489 (1954).

DUCKERT, F., FLÜCKIGER, P., MATTER, M., KOLLER, F.: Clotting factor X. Physiologic and physico-chemical properties. Proc. Soc. exp. Biol. (N.Y.) 90, 17 (1955).

DUCKERT, F., JUNG, Y., SCHMERLING, D. H.: A hitherto undescribed congenital hemorrhagic diathesis, probably due to fibrin stabilizing factor deficiency. Thrombos. Diathes. haemorrh. (Stuttg.) 5, 179 (1961).

Duckert, F., Koller, F., Matter, M.: Purification and physiological properties of factor VII from plasma and serum. Separation from prothrombin. Proc. Soc. exp. Biol. (N.Y.) **82**, 259 (1953).

Duckert, R.: The prephase accelerator. Present status. Thrombos. Diathes. haemorrh. (Stuttg.) **6**, 254 (1961).

Eagle, H.: Studies on blood coagulation. I. The role of prothrombin and of platelets in the formation of thrombin. J. gen. Physiol. **18**, 531 (1935).

Edman, P.: Method for determination of the amino acid sequence in peptides. Acta chem. scand. **4**, 283 (1950).

Egeberg, O.: A family with antihemophilic C factor (AHC = plasma thromboplastin antecedent) deficiency without bleeding tendency. Scand. J. clin. Lab. Invest. **14**, 478 (1962).

Egeberg, O.: Inherited antithrombin deficiency causing thrombophilia. Thrombos. Diathes. haemorrh. (Stuttg.) **13**, 576 (1965a).

Egeberg, O.: On the natural blood coagulation inhibitor system. Investigations of inhibitor factors based on antithrombin deficient blood. Thrombos. Diathes. haemorrh. (Stuttg.) **14**, 473 (1965b).

Egeberg, O.: Inherited fibrinogen abnormality causing thrombophilia. Thrombos. Diathes. haemorrh. (Stuttg.) **17**, 176 (1967).

Egeberg, O.: New families with hereditary hemorrhagic trait due to deficiency of fibrin stabilizing factor (fXIII). Thrombos. Diathes. haemorrh. (Stuttg.) **20**, 534 (1968).

Egeberg, O.: Factor XII defect and hemorrhage. Evidence for a new type of hereditary hemostatic disorder. Thrombos. Diathes. haemorrh. (Stuttg.) **23**, 432 (1970).

Eisen, V.: Fibrinolysis and the formation of biologically active polypeptides. Brit. med. Bull. **20**, 205 (1964).

Esnouf, M. P.: Plenary Session Papers, 12th Intern. Soc. Hematol., New York, 1968, p. 315.

Esnouf, M. P., Jobin, F.: The isolation of factor V from bovine plasma. Biochem. J. **102**, 660 (1967).

Esnouf, M. P., Williams, W. J.: The isolation and purification of a bovine plasma protein which is a substrate for the coagulant fraction of Russell's viper venom. Biochem. J. **84**, 62 (1962).

Evensen, S. A., Hjort, P. F.: Pathogenesis of disseminated intravascular coagulation. Plenary Session of XII. International Congr. of Hematology, 1970, p. 109.

Evensen, S. A., Jeremic, M.: Intravascular coagulation with generalized Shwartzman reaction induced by Liquoid: lack of protection by extreme thrombocytopenia. Thrombos. Diathes. haemorrh. (Stuttg.) **19**, 556 (1968).

Evensen, S. A., Jeremic, M.: Platelets and the triggering of intravascular coagulation. Brit. J. Haemat. **19**, 33 (1970).

Evensen, S. A., Jeremic, M., Hjort, P. F.: Intravascular coagulation with generalized Shwartzman reaction induced by a heparin-like anticoagulant (Liquoid). Thrombos. Diathes. haemorrh. (Stuttg.) **18**, 24 (1967).

Fantl, P., Sawers, R. J.: Anticoagulant specificity and physiologically inactive beta-prothromboplastin. Nature (Lond.) **177**, 1233 (1956).

Fantl, P., Sawers, R. J., Marr, A. G.: Investigation of a hemorrhagic disease due to beta-prothromboplastin deficiency complicated by a specific inhibitor of thromboplastin formation. Aust. Ann. Med. **5**, 163 (1956).

Feinstein, D. I., Chong, M. N. Y., Kasper, C. K., Rapaport, S. I.: Hemophilia A: Polymorphism detectable by a factor VIII antibody. Science **163**, 1071 (1969).

Fischer, Sh., Schwartz, M., Gottlieb, A., Ben Yoseph, N., Shapiro, S.: Fibrinolysis, fibrinogen and factor XIII in newborn infants. Thrombos. Diathes. haemorrh. (Stuttg.) **20**, 542 (1968).

Ganrot, P. O., Niléhn, J.-E.: Plasma prothrombin during treatment with dicumarol. II. Demonstration of an abnormal prothrombin fraction. Scand. J. clin. Lab. Invest. **22**, 23 (1968).

Ganrot, P. O., Niléhn, J. E.: Electrophoretic separation of two thrombin inhibitors in plasma and serum. Scand. J. clin. Lab. Invest. **24**, 11 (1969).

Gaston, L. W.: Studies on a family with elevated plasma level of factor V (proaccelerin) and a tendency to thrombosis. Pediatrics **68**, 376 (1966).

Gerhold, W. M., Tiongson, T., Mandel, E. E.: Studies of fibrin-stabilizing factor. Fed. Proc. **25**, 446 (1966).

GLADHAUG, A., PRYDZ, H.: Purification of the coagulation factors VII and X from human serum. Some properties of factor VII. Biochim. biophys. Acta (Amst.) 215, 105 (1970).

GLUECK, H. I., ROEHLL, W., JR.: Myocardial infarction in a patient with a Hageman (factor XII) defect. Ann. intern. Med. 64, 390 (1966).

GOBBI, F.: The fractionation properties of human factor VIII. Thrombos. Diathes. haemorrh. (Stuttg.) 4, 253 (1960).

GOODNIGHT, S. H., FEINSTEIN, D. I., ØSTERUD, B., RAPAPORT, S. I.: Factor VII antibodyneutralizing material in hereditary and acquired factor VII deficiency. Blood 28, 1 (1971).

GRAHAM, J. B., BARROW, B. M., HOUGIE, S.: Stuart clotting defect. II. Genetic aspect of a "new" haemorrhagic state. J. clin. Invest. 36, 492 (1957).

GRAHAM, R. C., JR., EBERT, R. H., RATNOFF, O. D., MOSES, J. M.: Pathogenesis of inflammation. II. In vivo observations of the inflammatory effects of activated Hageman factor and bradykinin. J. exp. Med. 121, 807 (1965).

GRAMMENS, G. L., PRASAD, A. S., MAMMEN, E. F., BARNHART, M. J.: Physico-chemical and immunological properties of bovine Hageman factor. Thrombos. Diathes. haemorrh. (Stuttg.) 25, 405 (1971).

GRAY, E. J., SCHAEFER, E. H., JENSEN, H.: Studies on role of accelerator factor in blood clotting mechanism. Acta haemat. (Basel) 15, 314 (1956).

HAANEN, C., HOMMES, F., MORSELT, G.: Some observations on the role of Hageman factor in blood coagulation. Thrombos. Diathes. haemorrh. (Stuttg.) 6, 261 (1961).

HAANEN, C., MORSELT, G., SCHOENMAKERS, J.: Contact activation of Hageman factor and the interaction of Hageman factor and plasma thromboplastin antecedent. Thrombos. Diathes. haemorrh. (Stuttg.) 17, 307 (1967).

HABERMANN, E.: In: Neue Aspekte der Trasylol-Therapie. Hrsg. R. GROSS u. G. KRONEBERG, S. 126. Stuttgart: Schattauer 1966.

HALL, C. E., SLAYTER, H. S.: The fibrinogen molecule: its size, shape and mode of polymerization. J. biophys. biochem. Cytol. 5, 11 (1959).

HAMMARSTEN, O.: Über das Fibrinogen. Pflügers Arch. ges. Physiol. 19, 563 (1879).

HAMMARSTEN, O.: Weitere Beiträge zur Kenntnis der Fibrinbildung. Z. phys. Chem. 28, 98 (1899).

HARDAWAY, R. M.: Syndromes of disseminated intravascular coagulation. With special reference to shock and hemorrhage, p. 484. Springfield: Charles C. Thomas 1966.

HARDISTY, R. M.: The reaction of blood coagulation factors with brain extracts. Brit. J. Haemat. 1, 323 (1955).

HARDISTY, R. M., MARGOLIS, J.: The role of Hageman factor in the initiation of blood coagulation. Brit. J. Haemat. 5, 203 (1959).

HARMISON, C. R., LANDABURU, R. H., SEEGERS, W. H.: Some physico-chemical properties of bovine thrombin. J. biol. Chem. 236, 1693 (1961).

HARMISON, C. R., SEEGERS, W. H.: Some physicochemical properties of bovine autoprothrombin. II. J. biol. Chem. 237, 3074 (1962).

HATHAWAY, W. E., ALSEVER, J.: The relation of the Fletcher factor to factor XI and XII. Brit. J. Haemat. 18, 161 (1970).

HECHT, E.: Lipids in blood clotting, Springfield, Ill.: Charles C. Thomas 1965.

HECHT, E., CHO, M. G., SEEGERS, W. H.: Thromboplastin: nomenclature and preparation of protein-free material different from platelet factor 3 or lipid activator. Amer. J. Physiol. 193, 584 (1958).

HECHT, E., SLOTTA, K. H.: The chemical nature of the lipid activator in blood coagulation. Amer. J. clin. Path. 37, 126 (1962).

HEMKER, H. C.: Preprothrombin (Complex?), a circulating anticoagulant in coumarin treated and vitamin K deficient patients. Thrombos. Diathes. haemorrh. (Stuttg.), Suppl. 13, 380 (1964).

HEMKER, H. C., ESNOUF, M. P., HEMKER, P. W., SWART, A. C. W., MACFARLANE, R. G.: Formation of prothrombin converting activity. Nature (Lond.) 215, 248 (1967).

HEMKER, H. C., KAHN, M. J. P.: Reaction sequence of blood coagulation. Nature (Lond.) 215, 1201 (1967).

HEMKER, H. C., KAHN, M. J. P., DEVILEE, P. P.: The adsorption of coagulation factors onto phospholipids. Its role in the reaction mechanism of blood coagulation Thrombos. Diathes. haemorrh. (Stuttg.) 24, 214 (1970).

Hemker, H. C., Loeliger, E. A., Veltkamp, J. J., eds.: Human blood coagulation, p. 397. Boerhaave series. Leiden: Univ. Press 1969.

Hemker, H. C., Muller, A. D.: Kinetic aspects of the interaction of blood clotting enzymes. VI. Localization of the site of blood-coagulation inhibitor by the protein induced by vitamin K absence (PIVKA). Thrombos. Diathes. haemorrh. (Stuttg.) 20, 78 (1968).

Hemker, H. C., Veltkamp, J. J., Hensen, A., Loeliger, E. A.: On the nature of prothrombin biosynthesis. Nature (Lond.) 200, 589 (1963).

Hemker, H. C., Veltkamp, J. J., Loeliger, E. A.: Kinetic aspects of the interaction of blood clotting enzymes. III. Demonstration of an inhibitor of prothrombin conversion in vitamin K deficiency. Thrombos. Diathes. haemorrh. (Stuttg.) 19, 346 (1968).

Henschen, A.: Number and reactivity of disulfide bonds in fibrinogen and fibrin. Arkiv Kemi 22, 355 (1964).

Henschen, A., Blombäck, B.: Amino acid composition of human and bovine fibrinogen and fibrin. Arkiv Kemi 23, 347 (1964).

Hjort, P.: Intermediate reactions in the coagulation of blood with tissue thromboplastin, convertin, accelerin, prothrombinase, p. 183. Thesis. Oslo: University Press 1957. Scand. J. clin. Lab. Invest. 9 (Suppl. 27), 183 (1957).

Hjort, P., Rapaport, S. I., Owren, P. A.: Platelet accelerator: identical to proaccelerin and adsorbed from plasma? Scand. J. clin. Lab. Invest. 7, 97 (1955).

Hoak, J. C., Swanson, L. W., Warner, E. D., Connor, W. E.: Myocardial infarction associated with severe factor XII deficiency. Lancet 1966 II, 884.

Högenauer, E., Lechner, K., Deutsch, E.: Isolation and characterization of blood clotting factors VII and X. Thrombos. Diathes. haemorrh. (Stuttg.) 19, 304 (1968).

Holemans, R., Roberts, H. R.: Hageman factor and in vivo activation of fibrinolysis. J. Lab. clin. Med. 64, 778 (1964).

Holmsen, H., Day, H. J., Stormorken, H.: The blood platelet release reaction. Scand. J. Haemat. 6 (Suppl. 8), 1 (1969).

Hopff, F.: Über die Hämophilie oder die erbliche Anlage zu tödlichen Blutungen. Inaugural-Dissertation Würzburg 1828. Quoted by Bulloch and Fildes, 1911.

Horn, R. G., Collins, R. D.: Studies on the pathogenesis of the generalized Shwartzman reaction. The role of granulocytes. Lab. Invest. 18, 101 (1968).

Hougie, C.: Effect of Russell's viper venom (Stypven) on Stuart clotting defect. Proc. Soc. exp. Biol. (N.Y.) 93, 570 (1956).

Hougie, C.: The role of factor V in the formation of blood thromboplastin. J. Lab. clin. Med. 50, 61 (1957).

Hougie, C.: Reactions of Stuart factor and factor VII with brain and factor V. Proc. Soc. exp. Biol. (N.Y.) 101, 132 (1959).

Hougie, C., Barrow, E. M., Graham, J. B.: Stuart clotting defect. I. Segregation of an hereditary hemorrhagic state from the heterogeneous group heretofore called "stable factor" (SPCA, proconvertin, factor VII) deficiency. J. clin. Invest. 36, 485 (1957).

Hougie, C., Denson, K. W. E., Biggs, R.: A study of the reaction product of factor VIII and factor IX by gel filtration. Thrombos. Diathes. haemorrh. (Stuttg.) 18, 211 (1967).

Hougie, C., Twomey, J. J.: Hemophilia Bm. A new type of factor IX deficiency. Lancet 1967 I, 698.

Howell, W. H.: The nature and action of the thromboplastic (zymoplastic) substance of the tissues. Amer. J. Physiol. 31, 1 (1912).

Howell, W. H.: The coagulation of blood. Harvey Lect. 12, 272 (1916–1917).

Hoyer, L. W., Breckenridge, R. T.: Immunologic studies of antihemophilic factor (AHF, factor VIII).: Cross-reacting material in a genetic variant of hemophilia A. Blood 32, 962 (1968).

Huseby, R. M., Murray, M.: Molecular structure of fibrinogen. I. Helical content and the role of the tyrosine moiety in the fibrinogen molecule. Biochim. biophys. Acta (Amst.) 133, 243 (1967a).

Huseby, R. M., Murray, M.: The circular dichroism and optical rotatory dispersion of fibrinogen: states of the aromatic amino acid residues in the molecule. Fed. Proc. 26, 537 (1967b).

HUSSAIN, Q. Z., NEWCOMB, T. F.: Effects of thrombin on factor V. Ann. Biochem. exp. Med. 23, 569 (1963)

HUTH, K., LÖFFLER, H., LECHELMAYR, U.: Verbrauchskoagulopathie bei unreifzelligen Leukosen. Verh. dtsch. Ges. inn. Med. 74, 147 (1968).

HVATUM, N., PRYDZ, H.: Studies on tissue thromboplastin. Its splitting into two separable parts. Thrombos. Diathes. haemorrh. (Stuttg.) 21, 217 (1969).

IATRIDIS, S. G., FERGUSON, J. H.: Effect of surface and Hageman factor on the endogenous or spontaneous activation of the fibrinolytic system. Thrombos. Diathes. haemorrh. (Stuttg.) 6, 411 (1961).

IATRIDIS, S. G., FERGUSON, J. H.: Active Hageman factor: A plasma lysokinase of the human fibrinolytic system. J. clin. Invest. 41, 1277 (1962).

INGRAM, C. Y. C., McBRIEN, D. J., SPENCER, H.: Fatal pulmonary embolism in congenital fibrinogenopenia. Acta haemat. (Basel) 35, 36 (1966).

IZAK, G., GALEWSKY, K.: Studies on experimentally induced hypercoagulable state in rabbits. Thrombos. Diathes. haemorrh. (Stuttg.) 16, 228 (1966).

JACKSON, C. M., HANAHAN, D. J.: Studies on bovine factor X. II. Characterization of purified factor X. Observations on some alterations in zone electrophoretic and chromatographic behavior occurring during purification. Biochemistry 7, 4506 (1968).

JACKSON, C. M., JOHNSON, T. F., HANAHAN, D. J.: Studies on bovine factor X. I. Large scale purification of the bovine plasma protein possessing factor X activity. Biochemistry 7, 4492 (1968).

JACKSON, D. P., BECK, E. A., CHARACHE, P.: Congenital disorders of fibrinogen. Fed. Proc. 24, 816 (1965).

JAQUES, L. B., FIDLAR, E., FELDSTED, E. T., MACDONALD, A. G.: Silicones and blood coagulation. Canad. med. Ass. J. 55, 26 (1946).

JENSEN, H., GRAY, E. J., SCHAEFER, E. H.: Formation of blood thrombokinase. Acta haemat. (Basel) 13, 377 (1955).

JOBIN, F., ESNOUF, M. P.: Studies on the formation of the prothrombin converting complex. Biochem. J. 102, 666 (1967).

JOHNSON, A. J., NEWMAN, J., HOWELL, M. B., PUSZKIN, S.: Purification of antihemophilic factor (AHF) for clinical and experimental use. Thrombos. Diathes. haemorrh. (Stuttg.), Suppl. 26, 377 (1967).

JOHNSON, P., MIHALYI, E.: Physicochemical studies of bovine fibrinogen. I. Molecular weight and hydrodynamic properties of fibrinogen and fibrinogen cleaned by sulfite in 5 M guanidine-HCl solution. Biochim. biophys. Acta (Amst.) 102, 467 (1965).

JOHNSON, S. A., SEEGERS, W. H.: Studies on the plasma defect in hemophilia. Rev. Hémat. 9, 529 (1954).

JORPES, J. E.: One hundred years of research on blood coagulation leading to the present day anticoagulant therapy in thrombosis. Thrombosis and Embolism, I. International Conference, Basel, 1954, p. 23–30. Basel: Benno Schwabe & Co. 1954.

JOSSO, F., LAVERGNE, J. M., GOUAULT, M., PROU-WARTELLE, O., SOULIER, J. P.: Differents états moleculaires du facteur II (prothrombine). Leur étude à l'aide de la staphylocoagulase et traités par les antagonistes de la vitamine K. Thrombos. Diathes. haemorrh. (Stuttg.) 20, 88 (1968).

JOSSO, F., PROU-WARTELLE, O.: Interaction of tissue factor and factor VII at the earliest phase of coagulation. Thrombos. Diathes. haemorrh. (Stuttg.), Suppl. 17, 45 (1965).

JÜRGENS, J.: The significance of the Hageman-factor for the effect of wettable surface on thrombocytes. Thrombos. Diathes. haemorrh. (Stuttg.) 7, 48 (1962).

KAHN, M. J. P., BOURGAIN, R. H.: A simplified preparation procedure of factor X by paper electrophoresis. Hemostase 5, 413 (1965).

KAHN, M. J. P., HEMKER, H. C.: Studies on blood coagulation factor V. I. The interaction of salts of fatty acids and coagulation factors. Thrombos. Diathes. haemorrh. (Stuttg.) 22, 417 (1969).

KAHN, M. J. P., HEMKER, H. C.: Studies on blood coagulation factor V. V. Changes of molecular weight accompanying activation of factor V by thrombin and the procoagulant protein of Russell's viper venom. Thrombos. Diathes. haemorrh. (Stuttg.) 27, 25 (1972).

KATTLOVE, H. E., SHAPIRO, S. S., SPIVACK, M.: Hereditary prothrombin deficiency. New Engl. J. Med. 282, 57 (1970).

Kekwick, R. A., Mackay, M. E., Nance, M. H., Record, B. R.: The purification of human fibrinogen. Biochem. J. **60**, 671 (1955).

Kellermeyer, R. W., Breckenridge, R. T.: The inflammatory process in acute gouty arthritis. I. Activation of Hageman factor by sodium urate crystals. J. Lab. clin. Med. **65**, 307 (1965).

Kellermeyer, R. W., Graham, R. C.: Kinins—possible physiologic and pathologic roles in man. New Engl. J. Med. **249**, 754 (1968).

Kezdy, F. J., Lorand, L., Miller, K. D.: Titration of active centers in thrombin solutions. Standardization of the enzyme. Biochemistry **157**, 2302 (1965).

Kingdon, H. S., Davie, E. W., Ratnoff, O. D.: The reaction between activated plasma thromboplastin antecedent and diisopropylphosphofluoridate. Biochemistry **3**, 166 (1964).

Kjærheim, A., Hovig, T.: The ultrastructure of haemostatic blood platelet plugs in rat mesenterium. Thrombos. Diathes. haemorrh. (Stuttg.) **7**, 1 (1962).

Koller, F., Krüsi, G., Luchsinger, P.: Über eine besondere Form haemorrhagischer Diathese. Schweiz. med. Wschr. **80**, 1101 (1950).

Koller, F., Loeliger, A., Duckert, F.: Experiments on a new clotting factor (factor VII). Acta haemat. (Basel) **6**, 1 (1951).

Koller, F., Loeliger, A., Duckert, F.: Le facteur VII. Rev. Hémat. **7**, 156 (1952).

Korsan-Bengtsen, K., Hjort, P. F., Ygge, J.: Acquired factor X deficiency in a patient with amyloidosis. Thrombos. Diathes. haemorrh. (Stuttg.) **7**, 558 (1962).

Korsan-Bengtsen, K., Ygge, J.: Purification of human thrombin on a carboxymethyl-cellulose column. Scand. J. Lab. clin. Invest. **13**, 591 (1961).

Kowalski, E.: Fibrinogen derived inhibitors of blood coagulation. Thrombos. Diathes. haemorrh. (Stuttg.) **4**, Suppl. Blood Clotting factors, 211 (1959).

Laki, K., Lorand, L.: On the solubility of fibrin clots. Science **108**, 280 (1948).

Lamy, F., Waugh, D. F.: The physical changes of prothrombin under various experimental conditions. Thrombos. Diathes. haemorrh. (Stuttg.) **2**, 12 (1958).

Lanchantin, G. F., Friedman, J. A.: Isolation of human plasma prothrombin of high specific activity by gel giltration. Proc. Soc. exp. Biol. (N.Y.) **114**, 584 (1963).

Lanchantin, G. F., Hart, D. W., Friedman, J. A., Saavedra, N. V., Mehl, J. W.: Amino acid composition of human plasma prothrombin. J. biol. Chem. **243**, 5379 (1968).

Larrieu, M. J., Meyer, D.: Abnormal factor IX during anticoagulant treatment. Lancet **1970 II**, 1085.

Larrieu, M. J., Meyer, D.: Heterogeneity of factor VIII and factor IX variants. Thrombos. Diathes. haemorrh. (Stuttg.), Suppl. **43**, 11 (1971).

Lasch, H. G., Heene, D. L., Huth, K., Sandritter, W.: Pathophysiology, clinical manifestations and therapy of consumption-coagulopathy ("Verbrauchskoagulopathie"). Amer. J. Cardiol. **20**, 381 (1967).

Lechner, K.: In: Hämophilie. Thrombos. Diathes. haemorrh. (Stuttg.), Suppl. **37**, 79 (1969).

Lechner, K., Deutsch, E.: Activation of factor X. Thrombos. Diathes. haemorrh. (Stuttg.) **13**, 314 (1965).

Lechner, L.: Immune reactive factor IX in acquired factor IX deficiency. Thrombos. Diathes. haemorrh. (Stuttg.) **27**, 19 (1972).

Lewis, J. H.: Synthesis of AHF in von Willebrand's disease. Blood **23**, 233 (1964).

Lewis, J. H., Merchant, W. R.: The probable identity of glass factor with Hageman factor. J. clin. Invest. **37**, 911 (1958).

Lewis, J. H., Walters, D., Didisheim, P., Merchant, W. R.: Application of continuous flow electrophoresis to the study of the blood coagulation proteins and fibrinolytic enzyme system. I. Normal human materials. J. clin. Invest. **37**, 1323 (1958).

Lewis, M. L., Ware, A. G.: A simple procedure for separation of prothrombin and accelerator globulin from citrated human plasma. Proc. Soc. exp. Biol. (N.Y.) **84**, 636 (1953).

Lipinski, B., Worowski, K., Jeljaszewicz, J., Niewiarowski, S., Rejniak, L.: Participation of soluble fibrin monomer complexes and platelet factor 4 in the generalized Shwartzman reaction. Thrombos. Diathes. haemorrh. (Stuttg.) **20**, 285 (1968).

LISTER, J.: Croonian lecture: on the coagulation of the blood. Proc. roy. Soc. 12, 580 (1863).

LISTON, R.: Haemorrhagic idiosyncrasy. Lancet 1839 II, 137.

LOELIGER, E. A., HERS, J. F. P.: Chronic antithrombinaemia (antithrombin V) with hemorrhagic diathesis in a case of rheumatoid arthritis with hypergammaglobulinaemia. Thrombos, Diathes. haemorrh. (Stuttg.) 1, 499 (1957).

LOEWY, A. G., DAHLBERG, J. E., DORWART, W. V., WEBER, M., EISELE, J.: A transamidase mechanism for insoluble fibrin formation. Biochem. biophys. Res. Commun. 15, 77 (1964).

LOEWY, A. G., DUNATHAN, K., KRIEL, R., WOLFINGER, H. L.: Fibrinase. I. Purification of substrate and enzyme. J. biol. Chem. 236, 2625 (1961a).

LORAND, L.: "Fibrinopeptide". New aspects of the fibrinogen-fibrin transformation. Nature (Lond.) 167, 992 (1951).

LORAND, L.: Fibrino-peptide. Biochem. J. 52, 200 (1952).

LORAND, L., DOWNEY, J., GOTOH, T., JACOBSEN, A., TOKURA, S.: The transpeptidase system which crosslinks fibrin by γ-glutamyl-ε-lysine bonds. Biochem. biophys. Res. Commun. 31, 222 (1968).

LORAND, L., JACOBSEN, C. A.: Studies on the polymerisation of fibrin. The role of the globulin: Fibrin-stabilizing factor. J. biol. Chem. 230, 401 (1958).

LORAND, L., KONISHI, K.: Activation of fibrin-stabilizing factor of plasma by thrombin. Arch. Biochem. 105, 58 (1964).

LORAND, L., KONISHI, K.: Separation of activated fibrin-stabilizing factor from thrombin. Biochim. biophys. Acta (Amst.) 121, 177 (1966).

LORAND, L., KONISHI, K., JACOBSEN, A.: Transpeptidation mechanism in blood clotting. Nature (Lond.) 194, 1148 (1962).

LÜSCHER, E. F.: Kontaktivierung, biologisch gesehen. Thrombos. Diathes. haemorrh. (Stuttg.), Suppl. 25, 79 (1968).

LYTTLETON, J. W.: The antithrombin activity of human plasma. Biochem. J. 58, 8 (1954).

MACFARLANE, R. G.: The coagulant action of Russell's viper venom. The use of antivenom in defining its reaction with a serum factor. Brit. J. Haemat. 7, 496 (1961).

MACFARLANE, R. G.: In: Metabolism and physiological significance of lipids (R. M. C. DAWSON and D. N. RHODES, eds.), p. 325—335. London-New York 1964.

MACFARLANE, R. G.: An enzyme cascade in the blood clotting mechanism, and its function as a biochemical amplifier. Nature (Lond.) 202, 495 (1964).

MACFARLANE, R. G., BIGGS, R., ASH, B. J., DENSON, K. W. E.: The activation and consumption of factor X in recalcified plasma. The effect of added factor VIII and Russell's viper venom. Brit. J. Haemat. 10, 530 (1964).

MAGNUSSON, S.: NH_2-terminal amino acid residues in bovine prothrombin and in prothrombin activation mixtures. With a note on the amino acid composition of chromatographically purified bovine prothrombin. Arkiv Kemi 23, 271 (1965a).

MAGNUSSON, S.: Preparation and carbohydrate analysis of bovine prothrombin. Arkiv Kemi 23, 285 (1965b).

MAGNUSSON, S.: Fractionation of bovine prothrombin preparations by gradient chromatography on DEAE-cellulose columns. Arkiv Kemi 24, 217 (1965c).

MAGNUSSON, S.: Preparation of highly purified bovine thrombin (E.C.3.4.4.13) and determination of its N-terminal amino acid residues. Arkiv Kemi 24, 349 (1965d).

MAGNUSSON, S.: Purification of prothrombin from human citrated plasma fraction II + III (Cohn's method). Arkiv Kemi 24, 367 (1965e).

MAGNUSSON, S.: Edman degradation of components of the bovine and human prothrombin-thrombin systems. Arkiv Kemi 24, 375 (1965f).

MAGNUSSON, S.: On the structure of thrombin and prothrombin. In: Human blood coagulation, ed. HEMKER et al., p. 18. Leiden: University Press 1969.

MAMMEN, E. F.: Mechanism of factor VIII inactivation. Fed. Proc. 24, 452 (1965).

MAMMEN, E. F.: Physiology and biochemistry of blood coagulation. In: Thrombosis and bleeding disorders, p. 554, ed. BANG et al. New York: Academic Press 1971.

MAMMEN, E. F., RAMIEN, A.: Die Anwendung der Gelfiltration zur weiteren Reinigung von Rinder-Prothrombin. Thrombos. Diathes. haemorrh. (Stuttg.) 8, 37 (1962).

MAMMEN, E. F., THOMAS, W. R., SEEGERS, W. H.: Activation of purified prothrombin to autoprothrombin I or autoprothrombin II (platelet cofactor II) or autoprothrombin II-A. Thrombos. Diathes. haemorrh. (Stuttg.) 5, 218 (1960a).

Mammen, E. F., Yoshinari, M., Seegers, W. H.: Platelet cofactors as plasma components of the intrinsic blood clotting mechanism. Thrombos. Diathes. haemorrh. (Stuttg.) **5**, 38 (1960b).

Marciniak, E., Seegers, W. H.: Prethrombin as a new sub-unit of prothrombin. Nature (Lond.) **209**, 621 (1966).

Marcus, A. J., Ullman, H. L., Safier, L. B., Ballard, H. S.: Platelet phosphatids. Their fatty acid and aldehyd composition and activity in different clotting systems. J. clin. Invest. **41**, 2198 (1962).

Margaretten, W., Zunker, H. O., McKay, D. G.: Production of the generalized Shwartzman reaction in pregnant rats by intravenous infusion of thrombin. Lab. Invest. **13**, 552 (1964).

Margolis, J.: Initiation of blood coagulation by glass and related surfaces. J. Physiol. (Lond.) **137**, 95 (1957).

Margolis, J.: Activation of plasma by contact with glass. Evidence for a common reaction which releases plasma kinin and initiates coagulation. J. Physiol. (Lond.) **144**, 1 (1958a).

Margolis, J.: Activation of permeability factor in plasma by contact with glass. Nature (Lond.) **181**, 635 (1958b).

Margolis, J.: The role of Hageman factor in plasma/surface reactions. In: Hemophilia and other hemorrhagic states (K. M. Brinkhous, ed.), p. 208. Chapel Hill: University of North Carolina Press 1959.

Margolis, J.: The mode of action of Hageman factor in the release of plasma kinin. J. Physiol. (Lond.) **151**, 238 (1960).

Margolis, J.: The interrelationship of coagulation plasma and release of peptides. Ann. N.Y. Acad. Sci. **104**, 133 (1963).

Matacic, S., Loewy, A. G.: The identification of isopeptide crosslinks in insoluble fibrin. Biochem. biophys. Res. Commun. **30**, 356 (1968.

McGrath, J. M., Stewart, G. J.: The effects of endotoxin on vascular endothelium. J. exp. Med. **129**, 833 (1969).

McKay, D. G.: Disseminated intravascular coagulation: and intermediary mechanism of disease, p. 493. New York: Harper Row publ. 1965.

Mellanby, J.: The coagulation of the blood. J. Physiol. (Lond.) **38**, 28, 441 (1909).

Ménaché, D.: Constitutional and familial abnormal fibrinogen. In: Fibrinogen and fibrin. Turnover of clotting factors. Ed. Hunter et al. Thrombos. Diathes. haemorrh. (Stuttg.), Suppl. **13**, 193 (1964).

Ménaché, D.: Congenitally abnormal fibrinogens. Thrombos. Diathes. haemorrh. (Stuttg.), Suppl. **39**, 307 (1970).

Ménaché, D., Boivin, P.: Deficit acquis en facteur X chez une malade atteint d'amyloidose primitive: injection d'un fraction CSB. Nouv. Rev. franç. Hémat. **2**, 868 (1962).

Meyer, D., Lavergne, J.-M., Larrieu, M.-J., Josso, F.: Cross-reacting material in congenital factor VIII deficiencies. (Haemophilia A and von Willebrand's disease), p. 183. New York: Pergamon Press Inc. 1972.

Miller, L. L., Hanavan, H. R., Titthasiri, N., Chowdhury, A.: Dominant role of the liver in the biosynthesis of the plasma proteins with special reference to the plasma mucoproteins (seromucoid) ceruloplasmin and fibrinogen. Chemistry **44**, 17 (1964).

Miller, L. L. John, D. W.: Factors affecting the net fibrinogen biosynthesis by the isolated perfused rat liver. Thrombos. Diathes. haemorrh. (Stuttg.), Suppl. **34**, 127 (1970).

Miller, S. P.: Coagulation dynamics in factor V deficiency: A family study with a note on the occurrence of thrombophlebitis. Thrombos. Diathes. haemorrh. (Stuttg.) **13**, 500 (1965).

Miller, S. P., Siggerud, J.: Abnormal blood coagulation in carriers of haemophilia. J. Lab. clin. Med. **63**, 601 (1964).

Monkhouse, F. C., France, E. S., Seegers, H. W.: Studies on the antithrombin and heparin cofactor activities of a fraction adsorbed from plasma by aluminium hydroxide. Circulat. Res. **3**, 397 (1955).

Morawitz, P.: Beiträge zur Kenntnis der Blutgerinnung. Arch. klin. Med. **79**, 215 (1904).

Morawitz, P.: Die Chemie der Blutgerinnung. Ergebn. Physiol. **4**, 307 (1905).

MOSESSON, M. W.: Molecular heterogeneity of human fibrinogen. Thrombos. Diathes. haemorrh. (Stuttg.), Suppl. **34**, 63 (1970).

MOSESSON, M. W., ALKJAERSIG, N., SWEET, B., SHERRY, S.: Human fibrinogen of relatively high solubility. Comparative biophysical, biochemical, and biological studies with fibrinogen of lower solubility. Biochemistry **6**, 3279 (1967).

MOSESSON, M. W., SHERRY, S.: The preparation and properties of human fibrinogen of relatively high solubility. Biochemistry **5**, 2829 (1966).

MÜLLER-BERGHAUS, G., LASCH, H.-G.: Untersuchungen über Beziehungen zwischen Gefäß- und Gerinnungsfaktoren beim Sanarelli-Shwartzman-Phänomen. Thrombos. Diathes. haemorrh. (Stuttg.) **9**, 335 (1963).

MÜLLER-BERGHAUS, G., LASCH, H. G.: Consumption of Hageman factor activity in the generalized Shwartzman reaction induced by Liquoid. Its prevention by inhibition of Hageman factor activation. Thrombos. Diathes. haemorrh. (Stuttg.) **23**, 386 (1970).

MÜLLER-BERGHAUS, G., LASCH, H. G.: Hageman factor activity in liquoid-induced consumption coagulopathy. Thrombos. Diathes. haemorrh. (Stuttg.) **26**, 38 (1971).

MUIRHEAD, C. R., TRIANTAPHYLLOPOULOS, D. C.: Anticoagulants produced by thrombin from fibrin, the effect on blood coagulation, some physical characteristics. Thrombos. Diathes. haemorrh. (Stuttg.) **26**, 211 (1971).

MURANO, G.: Some biochemical aspects of the prothrombin complex. Ph.D. Dissertation, Wayne State University, Detroit 1968.

MURANO, G.: Purification of thrombin. In: Thrombosis and bleeding disorders, ed. BANG et al., p. 124. New York: Acad. Press 1971.

MURRAY, M.: Vasculokinase, a clotting substance from arteries. Amer. J. clin. Path. **36**, 500 (1961).

MUSTARD, J. F., MEDWAY, W., DOWNIE, H. G., ROSWELL, H. C.: Effects of intravenous phospholipid containing phosphatidyl serine on blood clotting with particular reference to the Russell's viper venom time. Nature (Lond.) **196**, 1063 (1962).

NEMERSON, Y.: The reaction between bovine brain tissue factor and factors VII and X. Biochemistry **5**, 601 (1966).

NEMERSON, Y., SPAET, T. H.: The activation of factor X by extracts of rabbit brain. Blood **23**, 657 (1964).

NEWMAN, J., JOHNSON, A. J., KARPATKIN, M. H., PUSZKIN, S.: Methods for the production of clinically effective intermediate and high purity factor VIII concentrates. Brit. J. Haemat. **21**, 1 (1971).

NIEMETZ, J., NOSSEL, H. L.: Method of purification and properties of anti-XIa inhibitor of contact product. Thrombos. Diathes. haemorrh. (Stuttg.) **17**, 335 (1967).

NIEWIAROWSKI, S., BANKOWSKI, E., FIEDORUK, T.: Adsorption of Hageman factor (factor XII) on collagen. Experientia (Basel) **20**, 367 (1964).

NIEWIAROWSKI, S., BANKOWSKI, E., ROGOWICKA, I.: Studies on the adsorption and activation of the Hageman factor (factor XII) by collagen and elastin. Thrombos. Diathes. haemorrh. (Stuttg.) **14**, 387 (1965).

NIEWIAROWSKI, S., KOWALSKI, E.: Formation of an antithrombin-like anticoagulant during proteolysis of fibrinogen. Bull. Acad. pol. Sci. Cl. 2 **5**, 169 (1957).

NIEWIAROWSKI, S., KOWALSKI, E.: Une nouvel anticoagulant dérivé du fibrinogene. Rev. Hémat. **13**, 320 (1958).

NIEWIAROWSKI, S., PROU-WARTELLE, O.: Role du facteur contact (Facteur Hageman) dans la fibrinolyse. Thrombos. Diathes. haemorrh. (Stuttg.) **3**, 593 (1959).

NIEWIAROWSKI, S., STACHURSKA, J., WEGRZYNOWICZ, Z.: Arginine esterase activity of the contact (Hageman) factor. Thrombos. Diathes. haemorrh. (Stuttg.) **7**, 514 (1962).

NILSSON, I. M., BLOMBÄCK, M.: Von Willebrand's disease in Sweden. Its pathogenesis and treatment. Acta med. scand. **164**, 263 (1959).

NILSSON, I. M., BLOMBÄCK, M.: Von Willebrand's disease in Sweden—occurrence, pathogenesis and treatment. Thrombos. Diathes. haemorrh. (Stuttg.) **9**, Suppl. 11, 103 (1963).

NILSSON, I. M., BLOMBÄCK, M., JORPES, E., BLOMBÄCK, B., JOHANSSON, S. A.: Von Willebrand's disease and its correction with human plasma fraction 1–0. Acta med. scand. **159**, 179 (1957).

NILSSON, I. M., THIELEN, A., VON FRANCKEU, I.: Carriers of haemophilia A. A laboratory study. Acta med. scand. **165**, 357 (1959).

Norman, J. C., Covelli, V. H., Sise, H. S.: Transplantation of the spleen: Experimental cure of hemophilia. Surgery **64**, 1 (1968).

Nossel, H. L.: The contact phase of blood coagulation, p. 176. Oxford: Blackwell Scientific Publication 1964.

Nossel, H. L.: Differential consumption of coagulation factors resulting from activation of the extrinsic (tissue thromboplastin) or the intrinsic (foreign surface contact) pathways. Blood **29**, 331 (1967).

Nossel, H. L.: The earliest stages of blood coagulation. In: Recent advances in blood coagulation, ed. Poller, p. 39. London: J. A. Churchill 1969.

Nossel, H. L., Niemetz, J.: A normal inhibitor of the blood coagulation contact reaction product. Blood **25**, 712 (1965).

Nossel, H. L., Rubin, H., Drillings, M., Hsieh, R.: Inhibition of Hageman factor activation. J. clin. Invest. **47**, 1172 (1968).

Nussbaum, M., Morse, B. S.: Plasma fibrin stabilizing factor activity in various diseases. Blood **23**, 669 (1964).

O'Brien, J. R.: The effect of some fatty acids and phospholipids on blood coagulation. Brit. J. exp. Path. **38**, 529 (1957).

Østerud, B., Berre, Å., Otnaess, A. B., Björklid, A., Prydz, H.: Activation of the coagulation factor VII by tissue thromboplastin and Calcium. Biochemistry **11**, 2853 (1972).

Øzge-Anwar, A. H., Movat, H. Z., Scott, J. G.: The kinin system of human plasma. IV. The interrelationship between the contact phase of blood coagulation and the plasma kinin system in man. Thrombos. Diathes. haemorrh. (Stuttg.) **27**, 139 (1972).

Ogston, D., Bennett, N. B., Ogston, C. M., Ratnoff, O. D.: The assay of a plasma component necessary for the generation of a plasminogen activator in the presence of Hageman factor (Hageman factor co-factor). Brit. J. Haemat. **20**, 209 (1971).

Ottaviani, P. F., Mándelli, F., Fontana, L., Morelli, R.: Comportomento del fattore stabilizzante fibrinico nelle epatopatie. Progr. Med. **21**, 115 (1965).

Owen, C. A., Bollman, J. L.: Prothrombin conversion factor of dicoumarol plasma. Proc. Soc. exp. Biol. (N.Y.) **67**, 231 (1948).

Owen, C. A., Magath, T. B., Bollman, J. L.: Prothrombin conversion factors in blood coagulation. Amer. J. Physiol. **166**, 1 (1951).

Owren, P. A.: Nye undersøkelser over blodets koagulasjon. Proc. Norwegian Acad. of Science p. 21 (1944).

Owren, P. A.: The coagulation of blood, investigations on a new clotting factor. Acta med. scand. **128** (Suppl. 194), 327 (1947a).

Owren, P. A.: The fifth coagulation factor (factor V). Preparation and properties. Biochem. J. **43**, 136 (1948).

Owren, P. A.: New factors concerned in the coagulation of blood. Bull. Acad. Suisse Sci. Med. **3**, 163 (1947/48b).

Owren, P. A.: The diagnostic and prognostic significance of plasma prothrombin and factor V levels in parenchymatous hepatitis and obstructive jaundice. Scand. J. clin. Lab. Invest. **1**, 131 (1949).

Owren, P. A.: The diagnosis of the hemophilic conductor. Proc. Int. Soc. Haemat. 1950. Paper no. 104A.

Owren, P. A.: Proconvertin—the new clotting factor. Scand. J. clin. Lab. Invest. **3**, 168 (1951).

Owren, P. A.: New clotting factors. In: Blood clotting and allied problems. 5th Conference, p. 92. New York: Josiah Macy Foundation 1952a.

Owren, P. A.: La proconvertine. Rev. Hémat. **7**, 147 (1952b).

Owren, P. A.: Prothrombin and accessory factors. Clinical significance. Amer. J. Med. **14**, 201 (1953).

Owren, P. A.: Thrombotest. A new method for controlling anticoagulant therapy. Lancet **1959 II**, 774.

Owren, P. A.: Critical study of tests for control of anticoagulant therapy. Thrombos. Diathes. (haemorrh.) Stuttg.) **7**, Suppl. 1, 294 (1962).

Owren, P. A.: Control of anticoagulant therapy. The use of new tests. Arch. intern. Med. **3**, 248 (1963).

Owren, P. A.: Recent advances in the control of anticoagulant therapy. Geriatrics **26**, 74 (1971).

PAPAHADJOPOULOS, D., HOUGIE, C., HANAHAN, D. J.: Influence of surface charge of phospholipids on their clot-promoting activity. Proc. Soc. exp. Biol. (N.Y.) **111**, 412 (1962).

PAPAHADJOPOULOS, D., HOUGIE, C., HANAHAN, D. J.: Purification of bovine factor V: a change of molecular size during blood coagulation. Biochemistry **3**, 264 (1964).

PAPAHADJOPOULOS, D., YIN, E. T., HANAHAN, D. J.: Purification and properties of bovine factor X. Molecular changes during activation. Biochemistry **3**, 1931 (1964).

PASCUZZI, C. A., SPITTEL, J. A., THOMPSON, J. H., OWEN, C. A.: Thromboplastin generation accelerator, a newly recognized component of the blood coagulation mechanism present in excess in certain thrombotic states. J. clin. Invest. **40**, 1006 (1961).

PATEK, A. J., TAYLOR, F. H. L.: Hemophilia. II. Some properties of a substance obtained from normal human plasma effective in accelerating the coagulation of hemophilic blood. J. clin. Invest. **16**, 113 (1937).

PAVLOVSKY, A.: Contribution to the pathogenesis of hemophilia. Blood **2**, 185 (1947).

PECHET, L., COCHIOS, F., DEYKIN, D.: Further studies on the "Dynia" clotting abnormality. Thrombos. Diathes. haemorrh. (Stuttg.) **17**, 365 (1967).

PECHET, L., KASTRUL, J. J.: Amyloidosis associated with factor X (Stuart) deficiency. Case report. Ann. intern. Med. **61**, 315 (1964).

PENICK, P. D., ROBERTS, H. R., DEJANOV, I. I.: Covert intravascular clotting. Fed. Proc. **24**, 285 (1965).

PFUELLER, S., SOMER, J. B., CASTALDI, P. A.: Haemophilia due to an abnormal factor IX. Coagulation **2**, 213 (1969).

PISANO, J. J., FINLAYSON, J. S., PEYTON, M. P.: Cross-link in fibrin polymerized by factor XIII: ε-(γ-glutamyl)lysine. Science **160**, 892 (1968).

PITLICK, F.: The significance of factor IX inhibition by heparin in intrinsic blood coagulation. Thesis, University of Washington (1968).

POOL, J. G.: Antihemophilic globulin (AHG, factor VIII) activity in spleen. Fed. Proc. **25**, 317 (1966).

POOL, J. G., HERSHGOLD, E. J., PAPPENHAGEN, A. R.: High-potency actihaemophilic factor concentrate prepared from cryoglobulin precipitate. Nature (Lond.) **203**, 312 (1964).

POOLE, J. C. F., ROBINSON, D. S.: Further observations on the effects of ethanolamine phosphatide on plasma coagulation. Quart. J. exp. Physiol. **41**, 295 (1956).

PRENTICE, C. R. M., RATNOOF, O. D., BRECKENRIDGE, R. T.: Experiments on the natur of the prothrombin-converting principle: alteration of proaccelerin by thrombin. Brit. J. Haemat. **13**, 898 (1967).

PRYDZ, H.: Studies on proconvertin (factor VII). II. Purification. Scand. J. clin. Lab. Invest. **16**, 101 (1964).

PRYDZ, H.: Some characteristics of purified factor VII preparations. Scand. J. clin. Lab. Invest. **17**, Suppl. 84, 78 (1965).

PRYDZ, H., GLADHAUG, Å.: Factor X. Immunological studies. Thrombos. Diathes. haemorrh. (Stuttg.) **25**, 157 (1971).

QUICK, A. J.: Calcium in the coagulation of the blood. Amer. J. Physiol. **131**, 455 (1940).

QUICK, A. J.: Hemorrhagic diseases and the physiology of hemostasis. Springfield, Illinois: Charles C. Thomas 1942.

RAPAPORT, S. I.: Possible relationships between clotting factors in vitro and intravascular clotting. Angiology **10**, 391 (1959).

RAPAPORT, S. I., AAS, K., OWREN, P. A.: Effect of glass upon activity of various plasma clotting factors. J. clin. Invest. **34**, 9 (1955).

RAPAPORT, S. I., AMES, S. B., MIKKELSEN, S.: The levels of antihemophilic globulin and proaccelerin in fresh and bank blood. Amer. J. clin. Path. **31**, 297 (1955).

RAPAPORT, S. I., HJORT, P. F., PATCH, M. J.: Further evidence that thrombin-activation of factor VIII is an essential step in intrinsic clotting. Scand. J. clin. Lab. Invest. **17**, Suppl. 84 (1965).

RAPAPORT, S. I., PROCTOR, R. R., PATCH, M. J., YETTRA, M.: The mode of inheritance of PTA deficiency: evidence for the existence of major PTA deficiency and minor PTA deficiency. Blood **18**, 149 (1961).

RAPAPORT, S. I., SCHIFFMAN, I., PATCH, M. J., AMES, S. B.: The importance of activation of antihemophilic globulin and proaccelerin by traces of thrombin in the generation of intrinsic prothrombinase activity. Blood **21**, 221 (1963).

Rapport, M. M.: Activation of phospholipid thromboplastin by lecithin. Nature (Lond.) **178**, 591 (1956).

Ratnoff, O. D.: The biology and pathology of the initial stages of blood coagulation. In: Progress in hematology, eds. Moore and Brown, vol. 5, p. 204. New York: Grune & Stratton 1966.

Ratnoff, O. D., Busse, R. J., Sheon, R. P.: The demise of John Hageman. New Engl. J. Med. **279**, 760 (1968).

Ratnoff, O. D., Colopy, J. E.: A familial haemorrhagic trait associated with a deficiency of a clot promoting fraction of plasma. J. clin. Invest. **34**, 602 (1955).

Ratnoff, O. D., Crum, J. D.: Activation of Hageman factor by solutions of ellagic acid. J. Lab. clin. Med. **63**, 359 (1964).

Ratnoff, O. D., Davie, E. W.: Activation of Christmas factor (factor IX) by activated plasma thromboplastin antecedent (activated factor IX). Biochemistry **1**, 677 (1962).

Ratnoff, O. D., Davie, E. W., Mallett, D. L.: Studies on the action of Hageman factor: evidence that activated Hageman factor in turn activates plasma thromboplastin antededent. J. clin. Invest. **40**, 803 (1961).

Ratnoff, O. D., Miles, A. A.: The induction of permeability-increasing activity in human plasma by activated Hageman factor. Brit. J. exp. Path. **45**, 328 (1964).

Ratnoff, O. D., Rosenblum, J. M.: Role of Hageman factor in the initiation of clotting by glass. Amer. J. Med. **25**, 160 (1958).

Robbins, K. C.: A study on the conversion of fibrinogen to fibrin. Amer. J. Physiol. **142**, 581 (1944).

Roberts, H. R., Grizzle, J. E., McLester, W. D., Penick, G. D.: Genetic variants of hemophilia B: Detection by means of a specific PTC inhibitor. J. clin. Invest. **47**, 360 (1968).

Robinson, A. J., Aggeler, P. M., McNicol, Sp., Douglas, A. S.: An atypical genetical haemorrhagic disease with increased concentration of a natural inhibitor of prothrombin consumption. Brit. J. Haemat. **13**, 150 (1967).

Rodriguez-Erdmann, F.: Studies on the pathogenesis of the generalized Shwartzman reaction. III. Trigger mechanism for the activation of the prothrombin molecule. Thrombos. Diathes. haemorrh. (Stuttg.) **12**, 471 (1964).

Roka, L., ed.: Generalisierte intravaskuläre Gerinnung. Thrombos. Diathes. haemorrh. (Stuttg.), Suppl. **50**, 272 (1971).

Rosenthal, R. L., Dreskin, O. H., Rosenthal, N.: New hemophilia-like disease caused by deficiency of a third plasma thromboplastin factor. Proc. Soc. exp. Biol. (N.Y.) **82**, 171 (1953).

Rosenthal, R. L., Dreskin, O. H., Rosenthal, N.: Plasma thromboplastin antecedent (P.T.A.) deficiency: Clinical, coagulation, therapeutic and hereditary aspects of a new hemophilia-like disease. Blood **10**, 120 (1955).

Rozenberg, F. D., Waugh, D. F.: Multiple bovine thrombin components. J. biol. Chem. **245**, 5049 (1970).

Schiffman, S., Rapaport, S. E., Patch, M. J.: Starch block electrophoresis of clotting factors. Clin. Res. **12**, 110 (1964).

Schiffman, S., Rapaport, S. I., Chong, M. M. Y.: The mandatory role of lipid in the interaction of factors 8 and 9. Proc. Soc. exp. Biol. (N.Y.) **123**, 736 (1966).

Schimpf, K.: In: Uterine Hämostase. Herz und Blutgerinnung. Thrombos. Diathes. haemorrh. (Stuttg.), Suppl. **44**, 21 (1971).

Schmidt, A.: Über den Faserstoff und die Ursachen seiner Gerinnung. Arch. Anat. Physiol. Lpz., p. 545 (1861).

Schmidt, A.: Neue Untersuchung über die Faserstoffgerinnung. Pflügers Arch. ges. Physiol. **6**, 413 (1872).

Schmidt, A.: Quoted by Morawitz, 1905. Zur Blutlehre, Leipzig 1892.

Schneider, Ch. L.: Disseminated intravascular coagulation: Thrombosis versus fibrination, in clinical disease states. Thrombos. Diathes. haemorrh. (Stuttg.), Suppl. **36**, 1 (1969).

Schoenmakers, J. G. G., Kurstjens, R. M., Haanen, C., Zilliken, F.: Purification of activated bovine Hageman factor. Thrombos. Diathes. haemorrh. (Stuttg.) **9**, 546 (1963).

Schoenmakers, J. G. G., Matze R., Haanen, C., Zilliken, F.: Hageman factor a novel sialoglycoprotein with esterase activity. Biochim. biophys. Acta (Amst.) **101**, 166 (1965).

SCHWARTZ, M. L., SALVATORE, V. P., HILL, R. L., MCKEE, P. A.: The subunit structures of human plasma and platelet factor XIII (Fibrin-stabilizing factor). J. biol. Chem. **246**, 5851 (1971).

SCHWICK, G., SCHULTZE, H. E.: Immunochemische Untersuchungen mit Prothrombin und Thrombin. Clin. chim. Acta **4**, 26 (1959).

SEAMAN, A. J., OWREN, P. A.: An asolectin adsorbed substrate for proaccelerin assay. J. clin. Invest. **35**, 145 (1956).

SEEGERS, W. H.: The purification of prothrombin and thrombin. Chemical properties of the purified preparations. J. biol. Chem. **136**, 103 (1940).

SEEGERS, W. H.: The purification of prothrombin. Record Chem. Progr. **13**, 143 (1952).

SEEGERS, W. H.: Prothrombin, p. 728. Cambridge, Mass.: Harvard University Press 1962.

SEEGERS, W. H.: In: Blood coagulation hemorrhage and thrombosis (L. M. TOCANTIS and C. A. KAZAL, eds.), p. 174. New York: Grune & Stratton 1964.

SEEGERS, W. H.: Prothrombin in enzymology, thrombosis and hemophilia, p. 181. Springfield, Ill.: Thomas 1967.

SEEGERS, W. H., COLE, E. R., AOKI, N., HARMISON, C. R.: Separation of autoprothrombin III from bovine prothrombin preparations. Canad. J. Biochem. **42**, 229 (1964b).

SEEGERS, W. H., COLE, E. R., HARMISON, C. R., MARCINIAK, E.: Purification and some properties of autoprothrombin C. Canad. J. Biochem. **41**, 1047 (1963).

SEEGERS, W. H., COLE, E. R., HARMISON, C. R., MONKHOUSE, F. C.: Neutralization of autoprothrombin C activity with antithrombin. Canad. J. Biochem. **42**, 359 (1964a).

SEEGERS, W. H., COLE, E. R., MARCINIAK, E.: Partial activation of purified prothrombin: Derivation of autoprothrombin I with use of autoprothrombin C. Thrombos. Diathes. haemorrh. (Stuttg.) **7**, 239 (1962).

SEEGERS, W. H., HEENE, D., MARCINIAK, E.: Activation of purified prothrombin in ammonium sulphate solutions: Purification of autoprothrombin C. Thrombos. Diathes. haemorrh. (Stuttg.) **15**, 1 (1966).

SEEGERS, W. H., HEENE, D., MARCINIAK, E., IVANOVIC, N., CALDWELL, M. J.: Sensitivity of thrombin and autoprothrombin C to selected enzyme inhibitors. Life Sci. **4**, 425 (1965b).

SEEGERS, W. H., LANDABURU, R. H., FENICHEL, R. L.: Isolation of platelet/co-factor I from plasma and some properties of the preparation. Amer. J. Physiol. **190**, 1 (1957).

SEEGERS, W. H., MARCINIAK, E.: Some activation characteristics of the prothrombin subunit of prothrombin. Life Sci. **4**, 1721 (1965).

SEEGERS, W. H., MARCINIAK, E., HEENE, D.: Enzyme basis of prothrombin activation (blood clotting) with an analysis of hemophilia B. Tex. Rep. Biol. Med. **23**, 675 (1965a).

SEEGERS, W. H., MARCINIAK, E., KIPJER, R. K., YASUNAGA, K.: Isolation and some properties of prethrombin and autoprothrombin III. Arch. Biochem. Biophys. **121**, 372 (1967).

SEEGERS, W. H., MURANO, G., MCCOY, L., MARCINIAK, E.: The coagulation of blood: preliminary survey of thrombin and autoprothrombin zymogen structure. Life Sci. **8**, 925 (1969).

SEEGERS, W. H., NIEFT, M. L., LOOMIS, E. C.: Note on the adsorption of thrombin on fibrin. Science **101**, 520 (1945).

SHAPIRO, S. S., CHODOSH, B. T., ARONSON, D. L.: Congenital factor X dysproteinemia: a structural disorder of coagulation factor X. J. clin. Invest. (abs.) **49**, 87a (1970).

SHAPIRO, S. S., MARTINEZ, S. J., HOLBURN, R. H.: Congenital dysprothrombinemia: an inherited structural disorder of human prothrombin. J. clin. Invest. **48**, 2251 (1969).

SHAPIRO, S. S., WAUGH, D. F.: The purification of human prothrombin. Thrombos. Diathes. haemorrh. (Stuttg.) **16**, 469 (1966).

SHARP, A. A.: Viscous metamorphosis of blood platelets: a study of the relationship to coagulation factors and fibrin formation. Brit. J. Haemat. **4**, 28 (1958).

SHAW, S., PEGRUM, G. D., FARTHING, C. P., WOLFF, S.: Assessment of factor VII and factor X. Thrombos. Diathes. haemorrh. (Stuttg.) **15**, 294 (1966).

SHERMAN, L. A., FLETCHER, A. P., SHERRY, S.: In vivo transformation between fibrinogen of varying ethanol solubilities. Fed. Proc. **27**, 693 (1968).

SHERRY, S., ALKJAERSIG, N. K., FLETCHER, A. P.: Observations on the spontaneous arginine and lysine esterase activity of human plasma and their relation to Hageman factor. Thrombos. Diathes. haemorrh. (Stuttg.), Suppl. **20**, 243 (1966).

Shigeharu, N., Takahashi, H., Koida, M., Suzuki, T.: Partial purification of bovine plasma kallikreinogen, its activation by the Hageman factor. Biochem. biophys. Res. Commun. 32, 644 (1968).

Shulman, S.: The size and shape of bovine fibrinogen. Studies on sedimentation, diffusion and viscosity. J. Amer. chem. Soc. 75, 5846 (1953).

Shulman, S., Landaburu, R. H., Seegers, W. H.: Biophysical studies on platelet cofactor I preparations. Thrombos. Diathes. haemorrh. (Stuttg.) 4, 336 (1960).

Shulman, I., Smith, C. H., Erlandson, M., Fort, E., Lee, R. E.: A familial hemorrhagic disease in males and females characterized by combined antihemophilic globulin deficiency and vascular abnormality. Pediatrics 18, 347 (1956).

Simonetti, C., Casillas, G., Pavlovsky, A.: Purification du facteur VIII antihémophilique (FAH). Hémostase 1, 57 (1961).

Silver, M. J., Turner, D. L., Rodalewicz, I., Giordano, N., Holburn, R., Herb, S. F., Luckly, F. E.: Evaluation of activity of phospholipids in blood coagulation in vitro. Thrombos. Diathes. haemorrh. (Stuttg.) 10, 164 (1963).

Soulier, J.-P., Menache, D., Steinbuch, M., Blatrix, C. H., Josso, F.: Preparation and clinical use of PPSB (factors II, VII, X and IX concentrate). Thrombos. Diathes. haemorrh. (Stuttg.), Suppl. 35, 69 (1969).

Soulier, J.-P., Prou-Wartelle, O.: Preparation of a serum fraction rich in convertin (VII), Stuart factor (X) and antihemophilic B factor (IX). Fraction B.S.C. Experimental study in the rabbit. Nouv. Rev. franç. Hémat. 2, 27 (1962).

Soulier, J.-P., Prou-Wartelle, O.: New data on Hageman factor and plasma thromboplastin antedecent. The role of contact in the initial phase of blood coagulation. Brit. J. Haemat. 6, 88 (1960).

Soulier, J.-P., Prou-Wartelle, O., Menache, D.: Caractères différentiels des facterus Hageman et P.T.A. role du contact dans la phase initiale de la coagulation. Rev. franç. Étud. clin. biol. 3, 263 (1958).

Soulier, J.-P., Prou-Wartelle, O., Menache, D.: Hageman trait and PTA deficiency. The role of contact of blood with glass. Brit. J. Haemat. 5, 121 (1959).

Speer, R. J., Ridgway, H., Hill, M. J.: Activated human Hageman factor. Thrombos. Diathes. haemorrh. (Stuttg.) 14, 1 (1965).

Spittel, J. A., Pascuzzi, C. A., Thompson, J. H., Jr., Owen, C. A., Jr.: Acceleration of early stages of coagulation in certain patients with occlusive arterial or venous diseases: use of a modified thromboplastin generation test to evaluate clot. Proc. Mayo Clin. 35, 37 (1960).

Stites, D. P., Hershgold, E. J., Perlman, J. D., Fudenberg, M. M.: Factor VIII detection by hemagglutination inhibition. Hemophilia A and von Willebrand's disease. Science 171, 196 (1971).

Stormorken, H.: Species differences of clotting factors in ox, dog, horse and man. Proaccelerin and accelerin. Acta physiol. scand. 39, 121 (1957).

Stormorken, H.: The defect in von Willebrand's disease. In: Physiology of hemostasis and thrombosis (Johnson, S. A., Seegers, W. H., eds.), p. 179. Springfield, Ill. C.C. Thomas 1967.

Straub, W., Duckert, F.: The formation of the extrinsic prothrombin activator. Thrombos. Diathes. haemorrh. (Stuttg.) 5, 402 (1961).

Stryer, L., Cohen, C., Langridge, R.: Axial period of fibrinogen and fibrin. Nature (Lond.) 197, 793 (1963).

Surgenor, D. M., Wilson, N. A., Henry, A. S.: Factor V from human plasma. Thrombos. Diathes. haemorrh. (Stuttg.) 5, 1 (1961).

Telfer, T. P., Denson, K. W., Wright, D. R.: A "new" coagulation defect. Brit. J. Haemat. 2, 308 (1956).

Thompson, A. R., Davie, E. W.: Affinity chromatography of thrombin. Biochim. biophys. Acta (Amst.) 250, 210 (1971).

Tishkoff, G. H., Pechet, L., Alexander, B.: Some biochemical and electrophoretic studies on purified prothrombin, factor VII and factor X. Blood 15, 778 (1960).

Tishkoff, G. H., Williams, L. C., Brown, D. M.: Preparation of highly purified prothrombin complex. I. Crystallization, biological activity, and molecular properties. J. biol. Chem. 243, 4151 (1968).

Triantaphyllopoulos, D. C., Chen, E., Triantaphyllopoulos, E.: Nature of inhibition of prothrombin/consumption by lysed fibrinogen. Brit. J. Haemat. 16, 589 (1969).

TROUP, S. B., REED, C. F., MARINETTI, G. V., SWISCHER, S. N.: Thromboplastin factors in platelets and red blood cells: observations on their chemical nature and function in in vitro coagulation. J. clin. Invest. **39**, 342 (1960).

TURNER, D. L., SILVER, M. J.: Total synthesis of unsaturated phosphatidyl-serine and its activity in clotting systems. Nature (Lond.) **200**, 370 (1963).

VELTKAMP, J. J., DRION, E. F., LOELIGER, E. A.: Detection of the carrier state in hereditary coagulation disorders. Thrombos. Diathes. haemorrh. (Stuttg.) **19**, 279, 403 (1968).

VELTKAMP, J. J., MEILOF, J., RAMMELTS, H. G., VAN DER VLERK, D., LOELIGER, E. A.: Another genetic variant of haemophilia B. Haemophilia B Leyden. Scand. J.Haemat. **7**, 82 (1970).

VELTKAMP, J. J., MUIS, H., MULLER, A. D., HEMKER, H. C., LOELIGER, E. A.: Additional evidence for the existence of a precursor molecule of the prothrombin complex in oral anticoagulation. Thrombos. Diathes. haemorrh. (Stuttg.) **25**, 312 (1971).

VERSTRAETE, M., VERMYLEN, J., DONATI, M. B., eds.: Fibrinogen degradation products. Scand. J. Haemat., Suppl. **13**, 392 (1971).

VINAZZER, H., REINHARDT, F.: On the nature of anti-thrombin V. Thrombos. Diathes. haemorrh. (Stuttg.) **20**, 234 (1968).

VROMAN, L.: To be added in proof. Thesis, University of Utrecht, 1958.

VROMAN, L.: Behavior of coagulation factors at interfaces. In: Biophysical mechanism in vascular homeostasis and intravascular thrombosis (P. N. SAWYER, ed.), p. 81. New York: Appleton 1965.

VROMAN, L.: Surface activity in blood coagulation. In: Blood clotting enzymology (W. H. SEEGERS, ed.), p. 279. New York: Academic Press 1967.

VROMAN, L.: Biological aspects of surface activation. Thrombos. Diathes. haemorrh. (Stuttg.), Suppl. **25**, 89 (1968).

WAALER, B. A.: Contact activation in the intrinsic blood clotting system. Studies on a plasma product formed on contact with glass and similar surfaces. Thesis. Oslo University Press, 1959, p. 133. Scand. J. clin. Lab. Invest. **11** (Suppl. 37) (1959).

WALLS, W. D., LOSOWSKY, M. S.: Congenital deficiency of fibrin stabilizing factor. Coagulation **1**, III (1968).

WALSH, P. N.: The role of platelets in the contact phase of blood coagulation. Brit. J. Haemat. **22**, 237 (1972).

WARE, A. G., MURPHY, R. C., SEEGERS, W. H.: The function of Ac-globulin in clotting. Science **106**, 618 (1947).

WARE, A. G., SEEGERS, W. H.: Plasma accelerator globulin: partial purification, quantitative determination and properties. J. biol. Chem. **72**, 699 (1948).

WESSLER, S., REIMER, S. M., BLOEDE, M., NICKLES, M., SZLAT, B. J.: The role of human coagulation factors in serum-induced thrombosis. J. clin. Invest. **39**, 262 (1960).

WILLIAMS, W. J., ESNOUF, M. P.: The fractionation of Russell's viper (Vipera russellii) venom with special reference to the coagulant protein. Biochem. J. **84**, 52 (1962).

WILLIAMS, W. J., NORRIS, D. G.: Purification of a bovine plasma-protein factor (VII) which is required for the activity of lung microsomes in blood coagulation. J. biol. Chem. **241**, 1847 (1966).

WILNER, G. D., NOSSEL, H. L., DE ROY, C. C.: Activation of Hageman factor by collagen. J. clin. Invest. **47**, 2608 (1968).

WÖHLISCH, E.: Die Physiologie und Pathologie der Blutgerinnung. Ergebn. Physiol. **28**, 443 (1929).

WRIGHT, A. E.: On a method of determining the condition of blood coagulability for clinical and experimental purposes, and on the effect of the administration of calcium salts in haemophilia and actual or threatened haemorrhage. Brit. med. J. **1893 II**, 223.

YIN, E. T., WESSLER, S.: Heparin-accelerated inhibition of activated factor X by its natural plasma inhibitor. Biochim. biophys. Acta (Amst.) **201**, 387 (1970).

ZIMMERMAN, T. S., ARROYAVE, C. M., MÜLLER-EBERHARD, H. J.: A blood coagulation abnormality in rabbits deficient in the sixth component of complement (C 6) and its correction by purified C 6. J. exp. Med. **134**, 1591 (1971).

ZIMMERMAN, T. S., RATNOFF, O. D., POWELL, A. E.: Immunologic differentiation of classic hemophilia (Factor VIII deficiency) and von Willebrand's disease. J. clin. Invest. **50**, 244 (1971).

Postjunctional Supersensitivity and Subsensitivity of Excitable Tissues to Drugs

William W. Fleming, Joseph J. McPhillips, and David P. Westfall*

Table of Contents

* Department of Pharmacology, West Virginia University Medical Center, Morgantown, West Virginia, USA.

I. Introduction

A. Historical Background and Definitions

Supersensitivity may be defined as the phenomenon in which the amount of a substance required to produce a given biological response is less than "normal", i.e. the dose-response curve is shifted to the left. In a few instances there may also be an increase in the maximum response to a drug. However, this increase in maximal response is not a regular occurrence in supersensitivity. Likewise, supersensitivity may, but need not, be associated with a change in the slope of the dose-response curve. Thus, the one consistent sign of supersensitivity is the shift to the left of the dose-response curve.

In 1855, BUDGE made a remarkable observation. He cut the preganglionic fibers (decentralization) leading to the left superior cervical ganglion and the corresponding postganglionic fibers (denervation) on the right side of rabbits. Twenty-four to forty-eigth hours after the operation the pupils on the right side (the *denervated* side) were observed to be more dilated than the ones on the left (the *decentralized* side). This observation was investigated by others, especially LANGENDORFF (1900), who noted that a marked dilation of the denervated pupil occurred when a volatile anesthetic was administered to the animal. LANGENDORFF applied the term "paradoxical pupillary dilation" to describe this response. Not until the early part of this century was the cause of this paradoxical dilation understood to be an increased sensitivity of the denervated dilator muscle to substances (adrenaline and noradrenaline) released from the adrenal medulla.

The second example of supersensitivity grew out of an observation by PHILIPEAUX and VULPIAN (1863). The motor innervation of the tongue is via the hypoglossal nerve. Ordinarily, electrical stimulation of the lingual nerve does not affect the muscles of the tongue. However, beginning 4 to 6 days after the hypoglossal was cut, stimulation of the lingual nerve caused contraction of the tongue muscles. Again, after approximately 50 years and as the result of the work of several investigators, it was deduced that the denervated tongue had become supersensitive to acetylcholine. Enough acetylcholine could diffuse from the cholinergic vasodilator fibers of the lingual nerve to the supersensitive muscle to stimulate it. For a thorough historical review of supersensitivity the reader is referred to the excellent monograph by CANNON and ROSENBLUETH (1949).

Between 1939 and 1949, CANNON and his coworkers formulated a "Law of Denervation." As expressed by CANNON and ROSENBLUETH (1949), the law states: "When in a functional chain of neurons one of the elements is severed, the ensuing total or partial denervation of some of the subsequent elements in the chains causes a supersensitivity of all of the distal elements, including those not denervated, and effectors if present, to the excitatory or inhibitory action of chemical agents and nerve impulses; the supersensitivity is greater for the links which immediately follow the cut neurons and decreases progressively for more distal elements."

The monograph by CANNON and ROSENBLUETH (1949) demonstrates that the authors had an insight into the causes of supersensitivity which was very advanced in relation to the state of knowledge of cellular function which existed at that time. Some of the important points made by CANNON and ROSENBLUETH include the following. (1) There must be more than one mechanism underlying supersensitivity. (2) One of these mechanisms probably involves changes in the removal, by metabolism or storage, of normal transmitters. (3) In view of the very nonspecific nature of some examples of supersensitivity, another mechanism may involve a change in the physiological characteristics of the responding cells. They offer as a possibility an altered membrane permeability to ions such as calcium or potassium. As will be seen the conclusions of CANNON and his coworkers are still appropriate more than 20 years later.

The first tissue for which a mechanism for supersensitivity was delineated was skeletal muscle. This development was very thoroughly reviewed by THESLEFF (1960). Techniques combining intracellular electrical recording with localized iontophoretic application of acetylcholine established that, in innervated skeletal muscle, only the endplate responds to the transmitter. However, over a period of several days following denervation, the area of the muscle fiber sensitive to acetylcholine slowly spreads outward from the endplate and eventually includes the entire surface of the fiber. The sensitivity of the endplate itself does not change. The spread of sensitivity was specific for cholinergic agents but was not limited to substrates of cholinesterase. Local application of botulinum toxin, which chronically inhibits the release of acetylcholine, causes the same effect. A uniform sensitivity over the entire fiber surface is also a characteristic of fetal skeletal muscle.

In his review THESLEFF (1960) concluded: (1) The supersensitivity is due primarily to a spread of receptors outward from the endplate. He recognized, however, that other factors, including a decrease in cholinesterase and an increase in electrical excitability, probably contribute to the supersensitivity. (2) The supersensitive state probably represents the return to an original condition which precedes fully developed innervation of the muscle. (3) The supersensitivity is due to a loss of contact between the transmitter and the end organs and not to the degeneration of the nerves.

The electrophysiological evidence of a spread of receptors in supersensitive skeletal muscle has received further strong support from studies such as those by CREESE et al. (1971). These authors observed that the uptake of radioactively labelled decamethonium by skeletal muscle was increased 8–14 days after denervation and that the increase was primarily in the extrajunctional region. D-tubocurarine inhibited the uptake in both control and denervated preparations. It is the position of the authors of the present review that receptor spread has been established as the primary mechanism of supersensitivity in denervated skeletal muscle. Therefore, in considering supersensitivity in skeletal muscle in this review, emphasis will be placed on additional factors which may contribute to the phenomenon along with spread of receptors. It should be pointed out that MILEDI (1962, 1963) has questioned the conclusion that the neurotransmitter is the trophic factor in regulating receptor spread in skeletal muscle. This interesting question, however, appears not to be entirely settled and is beyond the scope of this review.

Supersensitivity and subsensitivity of salivary glands have been extensively studied by EMMELIN. For details, the reader is referred to the reviews by EMMELIN (1961, 1965). It will be seen that supersensitivity in exocrine glands shares all of the major attributes to be described later for supersensitivity in smooth muscle and cardiac muscle. These include: (1) Supersensitivity is induced by any procedure which chronically interrupts the normal contact between the neurotransmitter and the end organ. (2) The supersensitivity develops slowly. (3) It is nonspecific, the sensitivity being increased to both adrenergic and cholinergic agents, regardless of the means used to induce the supersensitivity.

EMMELIN presents evidence that the controlling factor in sensitivity is not the activity of the end organ but an effect of the transmitter itself (or a substance, such as pilocarpine, which mimics a transmitter) on the end organ. This is based on two observations. (1) The spontaneous release of transmitter in decentralized glands still maintains some degree of control over sensitivity even though there is no actual secretion of saliva being produced by the decentralized glands. (2) Sublingual glands, which secrete saliva spontaneously even in the absence of innervation, are capable of becoming equally as supersensitive as are submaxillary glands, which are inactive in the absence of innervation. These arguments hold, however, only if one considers the gross activity of the end organ as measurable, in this case, by salivary secretion. The critical "activity" in controlling sensitivity could conceivably be, for example, small changes in ion movement across the membrane in response to the transmitter.

Salivary glands seem to have a range of sensitivity which varies with the amount of transmitter to which they are exposed. An increase in the amount of transmitter reaching the end organ produces subsensitivity. This can be brought about by the chronic administration of a cholinesterase inhibitor. All of the sensitivity changes in salivary glands are readily reversible after the discontinuation of the procedure which induced the sensitivity change or after reinnervation.

TRENDELENBURG (1963 a) reviewed the subject of sensitivity to sympatho-mimetic amines. In his review some essential points were made regarding the measurement of sensitivity changes. This will be further considered sub-sequently (Section II). TRENDELENBURG'S review dealt with sensitivity changes in autonomically innervated structures with special emphasis on the nictitating membrane. Several important conclusions which are very relevant to the present review were made. Perhaps the most important is the clear demon-stration that there are at least two types of supersensitivity in peripheral structures.

One type is produced by cocaine. It is characterized by a high degree of specificity. The sensitivity is increased only to noradrenaline and a few other very closely related amines. It develops very rapidly. The other type of super-sensitivity is produced by such procedures as decentralization and chronic depletion of transmitter stores (i.e. by interruption of the normal contact between a transmitter and its end organ). This type of supersensitivity develops slowly and is very nonspecific. That is, the sensitivity is increased to agonists which are chemically and pharmacologically unrelated. Denervation (that is, postganglionic denervation) produces both types of supersensitivity. The con-trast between these two types of supersensitivity was further amplified sub-sequently (see, for example, TRENDELENBURG, 1963 b).

Several years later (TRENDELENBURG, 1966) the terms presynaptic and postsynaptic were introduced to differentiate these two types of supersensi-tivity. Presynaptic infers that some event, such as loss of uptake of nor-adrenaline, which alters the amount of agonist reaching the receptors of the end organ, is the cause of the apparent increase in sensitivity. Presynaptic supersensitivity is not a very satisfactory term since the sensitivity of respond-ing cells is not changed. Rather, there is an increase in the concentration of the active agent in the biophase with no change in the relation between the concentration in the biophase and the magnitude of response. Postsynaptic supersensitivity infers that the sensitivity of the responding cells is actually increased. Thus, postsynaptic supersensitivity is "true" supersensitivity since there is an increase in the magnitude of response to a given concentration of the active agent in the biophase.

TRENDELENBURG'S review (1963 a) also extensively discusses subsensitivity to indirectly acting sympathomimetics such as tyramine. This is produced by any procedure which either prevents the access of the amine to the endogenous stores of noradrenaline or depletes these stores. Such would therefore be con-sidered to be a presynaptic or prejunctional form of subsensitivity and is therefore not a subject of the present review.

There are some intriguing examples of supersensitivity in the central nervous system. STAVRAKY (1961) has described in detail many examples in which interruption of afferent pathways or surgical isolation of neurons produces supersensitivity to acetylcholine in both animals and man. In some instances this supersensitivity was shown to develop slowly and to be non-specific. That is, the sensitivity was increased to agents such as pentylene-tetrazol as well as to acetylcholine. These are the properties associated with postsynaptic (postjunctional) supersensitivity in peripheral structures (TREN-DELENBURG, 1963a, 1966). STAVRAKY also called attention to an increase in spontaneous activity and in the sensitivity to direct electrical stimulation of denervated neurons. As will be described later (Section III. C) these same changes have been described in association with postjunctional supersensitivity in certain peripheral tissues.

STAVRAKY (1961) has advanced supersensitivity as a possible mechanism underlying clinical epilepsy. He suggested that some pathological process may have resulted "in a partial isolation" of certain neurons. As a result these cells would become supersensitive to whatever stimuli they still receive and thus overreact, leading to convulsions. SHARPLESS (1964) has pointed out that chronic surgical isolation of cortical neurons leads to a greatly prolonged afterdischarge following repetitive stimulation. This characteristic develops slowly over several weeks and is likened to an epileptiform afterdischarge. However, SPEHLMANN (1971), using microelectrodes and microelectrophoretic techniques, was unable to find any evidence of supersensitivity to acetyl-choline associated with the epileptiform activity in chronically isolated cortex.

SHARPLESS (1964) has advocated the word *disuse* to indicate supersensi-tivity produced by various procedures which have as their common element the interruption of the normal stimulus input to an excitable cell. Thus the term disuse supersensitivity would be used for the same type of supersensitivity which TRENDELENBURG has called postsynaptic supersensitivity.

A recent and excellent review on supersensitivity in the central nervous system has been provided by SHARPLESS (1969). One particularly interesting finding reviewed by SHARPLESS is the development of supersensitivity of the temperature-regulating center of the hypothalamus produced by chronic ad-ministration of scopolamine. The supersensitivity was slow to develop and slowly reversible. The parallel between these results and the effect of chronic administration of atropine on peripheral structures (see EMMELIN, 1961, 1965) is striking.

SHARPLESS (1969) has discussed the intriguing possibility that the phenom-ena of tolerance and withdrawal reactions to CNS depressants may be a reflection of disuse supersensitivity. For example, the tendency for animals and man to undergo seizures is elevated for a period of time after withdrawal of chronically administered barbiturate. This could represent the development of supersensitivity in neurons as a response to the prolonged depressant effect of the barbiturate.

Consistent with this hypothesis are the results of DOMINIC and MOORE (1969) in studies of activity of animals centrally induced by indirectly acting sympathomimetics. They found that, as expected, α-methyltyrosine, an inhibitor of catecholamine synthesis, reduced the effect of amphetamine and ephedrine. However, as the treatment continued, the response to the amphetamine and ephedrine returned (i.e. "tolerance" developed). After the treatment was discontinued, there was supersensitivity to the amines. The authors concluded that, as the amount of transmitter decreased, supersensitivity of the postjunctional neurons increased to whatever amount of transmitter was released.

In 1965, GREEN and ROBSON suggested that tolerance to antihypertensive agents may involve supersensitivity. This is an important hypothesis which deserves more attention and research.

As with many incompletely understood phenomena, the terminology of supersensitivity can be confusing. As new evidence continually unfolds terms may cease to be completely appropriate. However, one must keep some historical perspective in the choice of terms. The topic chosen for this review is postjunctional supersensitivity and subsensitivity. Postjunctional sensitivity changes are defined as those due to an altered property of the responding cells themselves. Postjunctional supersensitivity is considered by us to be identical with postsynaptic supersensitivity (TRENDELENBURG, 1966) and disuse supersensitivity (SHARPLESS, 1964).

Our choice of postjunctional supersensitivity in preference to the other terms is based on the following. Some investigators consider the word synapse to represent a specific type of junction, i.e. that between two neurons. Some might incorrectly assume, therefore, that postsynaptic supersensitivity is a characteristic of neurons. The term disuse supersensitivity implies that inactivity of the end organ leads to the organ becoming supersensitive. Although this is an attractive theory, it is by no means established. See, for example, the arguments of EMMELIN (1965) in favor of contact between the neurotransmitter and the end organ being the controlling factor of the end organ's sensitivity.

By defining postjunctional supersensitivity as above, it is assumed that any type of supersensitivity which is not due to a change in the responsiveness of the target cells themselves may be considered prejunctional. This would include any mechanisms that alter the concentration of the agonist at site of action. Some investigators would object to this classification on the basis that some mechanisms which inactivate a drug may be located postjunctionally, i.e. within the responding cells. For example, under some conditions the enzyme catechol-O-methyl-transferase (COMT) located in smooth muscle cells may play a role in regulating the concentration of catecholamines at their receptors (KALSNER and NICKERSON, 1969). It could be argued that inhibition of COMT at a postjunctional site could produce supersensitivity to catecholamines. Such a mechanism would not be included in our definition of postjunctional supersensitivity. However, the importance of such a mechanism is yet to be established.

B. Scope and Goals of this Review

This review is restricted to postjunctional supersensitivity and subsensitivity. Thus, beyond discussing the two types of sensitivity change in Section I. A to establish their separate identities, there will be no further consideration of prejunctional supersensitivity or subsensitivity. As much as possible, discussion of experiments in which postjunctional sensitivity changes are complicated by the simultaneous presence of prejunctional sensitivity changes will be avoided.

All consideration of supersensitivity produced by cocaine will be avoided. It has been proposed that cocaine has a postjunctional as well as prejunctional effect (Maxwell et al., 1966; Kasuya and Goto, 1968). The relationship of this phenomenon to other types of postjunctional supersensitivity is not at all clear and must await further investigation.

The major emphasis will be on smooth muscle and cardiac muscle. This reflects the authors' feeling that these tissues incorporate the greatest amount of recent work on the subject of postjunctional supersensitivity and subsensitivity not covered in other reviews. Reference to other tissues will be made to bring in recent work and to allow contrasts and comparisons to smooth muscle and cardiac muscle. Particular attention will be paid to skeletal muscle in considering mechanisms of postjunctional supersensitivity because of the advanced state of knowledge concerning changes in that tissue brought about by denervation.

Sharpless (1969) has noted that investigators have made very few attempts to draw together concepts of supersensitivity in different types of organs and tissues. We strongly agree and hope that this review will make some small progress toward correcting that situation.

II. Quantitative Methods of the Study of Supersensitivity

Trendelenburg (1963a) has stressed the importance of the use of dose-response curves as opposed to single doses of drugs in estimating sensitivity changes. He also made it clear that quantitative measurements of supersensitivity must be based on the magnitude of horizontal shifts of dose-response curves. This is accomplished by comparing doses producing equal responses in control and experimental groups of preparations. Finally, Trendelenburg pointed out that the most accurate part of a dose-response curve from which to measure a shift is the steep portion of the curve, i.e. toward its center.

TRENDELENBURG (1965) and FOSTER (1966) have noted that equieffective doses of an agonist are normally distributed on a log scale, not an arithmetic scale. This matter is discussed in detail by FLEMING et al. (1972). The importance of this observation is that mean logs or geometric means of equieffective doses, not arithmetic means, are the appropriate and statistically valid estimate of the mean sensitivity in a single population of similar tissues. The geometric mean is the antilog of the mean log. It has the combined advantages of being based on the log normal distribution of the individual equieffective doses and having the same units as the individual doses.

A frequent error made in using geometric means is the presentation of the mean plus or minus a standard error. Since the logs of equieffective doses are normally distributed one can present the mean log $\pm$ S.E. However, since the variation about a mean log is equal on both sides, the variation about its antilog (i.e. the geometric mean) cannot possibly be. This problem can be circumvented by the presentation of the geometric mean with its 95 % confidence interval instead of a standard error. Specific examples of geometric means and confidence intervals can be seen in FLEMING et al. (1972).

The most accurate estimate of changes in sensitivity may be obtained by either of two methods. The first is by calculating the ratio of geometric means of equieffective doses from two groups of data, for example, from control and supersensitive tissues (TRENDELENBURG, 1965). The second method involves the determination of the difference between two mean logs (see for example, FOSTER, 1966; TRENDELENBURG, et al. 1969). This latter value may then be presented as the log shift of the dose-response curve. It should be noted that the log shift is equal to the log of the ratio of geometric means. A variation of the log-shift method of presenting the data is more appropriate in studies which are based on paired samples. The difference between the logs of equieffective doses in each pair is determined. The individual log differences in each pair are then averaged and treated statistically. Regardless of whether one chooses to present log shifts or ratios of geometric means, the statistical analyses such as t-tests or analyses of variance are performed on the log data.

A very great problem in some studies of sensitivity changes is a difference in baseline values between control and experimental groups. Differences in baseline values can affect the magnitude of responses. This is particularly true of heart rate and blood pressure. It can also be a factor in studies of smooth muscle. The resting tension applied to the intact nictitating membrane has a dramatic effect on responses and the determination of sensitivity (HAMPEL, 1934; WESTFALL et al., 1969).

A unique special case of this problem of baseline values is the existence of "tone" in the supersensitive nictitating membrane (LANGER, 1966). Tone results from the supersensitive nictitating membrane being maintained in a continual state of partial contraction in response to circulating levels of catecholamines which are generally too low to contract normal nictitating membranes. In the presence of tone, the sensitivity of the nictitating membrane to stimulant drugs is underestimated. LANGER (1966) and LANGER et al. (1967) give details of the methods of measuring and reducing tone. For the accurate quantitative determinations of changes in sensitivity, tone must be greatly reduced. However, the qualitative conclusions concerning supersensitivity based on experiments in the nictitating membrane in which tone was unsuspected (before 1966) are valid.

The importance of the slopes of dose-response curves to an analysis of supersensitivity has recently become recognized. For example, the slope of a dose-response curve obtained with noradrenaline in the nictitating membrane changes as the curve is moved in or out of the range of noradrenaline concentrations in which neuronal uptake is saturated (LANGER and TRENDELENBURG, 1969). Such a change in slope can create errors in estimates of the magnitude of the shift. Furthermore, postjunctional supersensitivity has been shown to alter the slope of the dose-response curve of methoxamine on the nictitating membrane (TRENDELENBURG,et al., 1970). Since methoxamine is not taken up into adrenergic neurons this particular change in slope cannot be related to the saturation of uptake.

WESTFALL (1970a) observed a remarkable inverse correlation between the slope of the dose-response curve to agonists in the vas deferens and the magnitude of the shift of the dose-response curve with the development of postjunctional supersensitivity. Similar correlations may exist in other tissues. However, accurate calculations of slopes have not generally been provided in studies of postjunctional supersensitivity. This may be a fruitful area for future study.

III. Analysis of Postjunctional Supersensitivity

A. Procedures which Produce Postjunctional Supersensitivity

The earliest methods used to induce this phenomenon were surgical denervation and decentralization. Adequate examples have already been given in Section I. A and in other reviews referred to therein. More recently many pharmacological means have become available and been widely used.

The first pharmacological tool to be widely adapted for the induction of postjunctional supersensitivity was the chronic administration of reserpine or other Rauwolfia alkaloids (BURN and RAND, 1958; FLEMING and TRENDELENBURG, 1961; EMMELIN, 1961). The supersensitivity must be considered to be the result of the chronic depletion of the transmitter (EMMELIN, 1961;

TRENDELENBURG and WEINER, 1962; FLEMING, 1963a). It is important to emphasize that reserpine produces a purely postjunctional type of supersensitivity. Reserpine is not able to cause prejunctional supersensitivity unless it is combined with an inhibitor of monoamine oxidase (GRAEFE et al., 1971; TRENDELENBURG, 1971).

There are problems associated with reserpine used as a tool to study supersensitivity. First, one must use enough reserpine to obtain nearly complete depletion. Second, large doses of reserpine render animals extremely ill. Of particular importance are a pronounced subnormal body temperature and diarrhea. As a consequence of diarrhea there may be alterations in water and electrolyte balance. Third, in addition to depleting noradrenaline stores in adrenergic neurons reserpine also has direct effects on effector cells. These effects are particularly apparent and serious in the heart.

WITHRINGTON and ZAIMIS (1961) and ZAIMIS (1961) have presented evidence that a single large dose of reserpine (1.0 mg/kg) produces signs of cardiac failure and cellular damage in the cat within 24 hours. Subsequently, WITHRINGTON and ZAIMIS (1967) presented evidence of some cardiac impairment in cats given a small daily dose of reserpine (0.01 mg/kg) for periods of several weeks. Cardiac depression by reserpine is probably a consequence of damage to metabolic pathways. SCHWARTZ and LEE (1960) found evidence of an uncoupling of oxidative phosphorylation in the hearts of cats and guinea pigs given reserpine, 5.0 mg/kg. Consistent with this observation is the fact that reserpine causes damage to mitochondria (WILCKEN et al., 1967; SUN et al., 1968; IWAYAMA et al., 1973). Fortunately, at least in the guinea pig heart, it is possible to separate the depletion of noradrenaline and the development of supersensitivity, on the one hand, from the cardiac depression and mitochondrial damage, on the other, by the proper choice of dose of reserpine, schedule of pretreatment and duration of pretreatment (WESTFALL and FLEMING, 1968a; IWAYAMA et al., 1973).

A metabolic depressant effect of reserpine has also been observed in the rat aorta (GILLIS, 1959). Reserpine (0.5 mg/kg/day for 10 days or one dose of 5.0 mg/kg) depressed oxygen consumption. Lower doses of reserpine did not have this effect. The doses required to produce depression of QO_2 in rat aorta exceed those which are optimal for producing supersensitivity in the rabbit aorta (HUDGINS and FLEMING, 1966).

It is interesting that reserpine has also been shown to induce biochemical changes in salivary glands. TAYLOR et al. (1967) have reported that reserpine caused a 4- to 5-fold increase in concentration of glycoprotein in the rat submaxillary gland. The peak effect was observed at 20–30 hours. This rapid time course and the fact that neither denervation nor decentralization caused a similar effect rules out any possibility that the increase in glycoprotein has any connection with supersensitivity.

Under some circumstances direct depressant effects of reserpine may mask supersensitivity. Westfall and Fleming (1968a) observed that 7 days of pretreatment of guinea pigs with reserpine, 0.1 mg/kg/day, resulted in supersensitivity of the isolated perfused heart. However, if the treatment were extended for 14 days, no supersensitivity was detectable. Other evidence from that study indicated that a direct depressant effect of reserpine is present after 14 days, but not 7 days, of administration of the same dose.

Extraneous effects of large doses of reserpine can also cause subcellular changes which may be incorrectly assumed to be associated with supersensitivity. Carrier and Shibata (1967) and Carrier et al. (1967) reported that reserpine caused supersensitivity of the rabbit aorta associated with marked reductions in aortic content of calcium, sodium and potassium. They concluded that the supersensitivity was a reflection of the altered ion content. However, later studies in the aorta (Hudgins and Harris, 1970; Garrett and Carrier, 1971) and the vas deferens (Westfall, 1970a, b) found supersensitivity produced by reserpine without any changes in total ion content. The explanation appears to lie in the use of higher doses of reserpine in the earlier studies (Garrett and Carrier, 1971).

The solution to this problem is (1) to use other methods to produce supersensitivity when possible or (2) to use the minimum effective dose of reserpine required to produce sufficient depletion and an optimal period of treatment to produce supersensitivity. The problem with (2) is that the dose and period vary with the tissue and the species. One must therefore determine the optimal conditions for each tissue. Fortunately, much of this information is in the literature. Examples are the very thorough studies of depleting effects of reserpine by Crout et al. (1962) and Lee (1967). A discussion of the vast literature on depletion by reserpine is beyond the scope of this review. However, if one does not find the necessary information in the literature, one must either measure catecholamine levels or use nerve stimulation or tyramine as an indicator of depletion and experiment with different doses and schedules of administering reserpine.

In recent years many different approaches to producing postjunctional supersensitivity have been investigated. The procedures which have been shown to produce the phenomenon may be classified as follows:

(1) Chronic blockade of receptors of the effector cells with drugs such as atropine (Emmelin, 1961) or scopolamine (Friedman et al., 1969).

(2) Chronic prevention of release of transmitter by:

(a) reserpine (transmitter depletion);

(b) adrenergic neuron blockers such as TM-10 (Trendelenburg and Weiner, 1962);

(c) botulinum toxin (Thesleff, 1960; Emmelin, 1961).

(3) Chronic surgical denervation or decentralization.

(4) Chemical denervation with 6-hydroxydopamine (HAEUSLER et al., 1969; MALING et al., 1971).

(5) Chronic ganglionic blockade (EMMELIN, 1961; TRENDELENBURG and WEINER, 1962; FLEMING, 1968).

(6) Alteration of sensory stimuli (BITO et al., 1971).

(7) Surgical interruption of *afferent* neural pathways (EMMELIN, 1965).

It will be recognized that some of these procedures are chosen on the basis of whether one wishes to interrupt adrenergic transmission (reserpine, neuron blockade, 6-hydroxydopamine) or cholinergic transmission (atropine, scopolamine, Botulinum toxin). In contrast, surgical interruption of nerve impulses, especially decentralization, and chronic ganglionic blockade are not limited to either adrenergic or cholinergic nerves.

Most of these procedures have their own characteristic disadvantages. Those of reserpine have already been extensively discussed. Surgical denervation or decentralization is often technically very difficult, particularly in the cardiovascular system. Atropine is long-lasting and makes it difficult to measure sensitivity changes to cholinergic agonists. Ganglionic blocking agents have rather short durations of action and must be administered two or three times per day. Botulinum toxin is very toxic systemically, must be applied locally and is hazardous to work with. 6-Hydroxydopamine causes degeneration of adrenergic neurons and thus prejunctional (WAGNER and TRENDELENBURG, 1971) as well as postjunctional supersensitivity.

It is logical to assume that all these procedures produce postjunctional supersensitivity by a common mechanism. All the pharmacological methods of producing the phenomenon share a common property with denervation and decentralization; they all chronically interrupt the normal contact between a neurotransmitter and its effector cells. In those tissues in which different procedures can and have been compared, such as the nictitating membrane and the vas deferens (see Section III. B), various procedures produce supersensitivity of similar nonspecificity, magnitude and time course. It seems logical to assume that all these procedures produce postjunctional supersensitivity because of the removal of some trophic influence of the transmitter or a reduction in the activity of the effector cells themselves (EMMELIN, 1961, 1965; TRENDELENBURG and WEINER, 1962).

An interesting observation should be noted in this section. DE LA LANDE et al. (1967) have found that 5-hydroxytryptamine acutely increases the sensitivity of the perfused artery of the rabbit ear. The sensitivity is nonspecifically increased to noradrenaline, histamine and angiotensin. The effect does not occur in depolarized arteries. Except for its immediate action, this effect of 5-HT has the characteristics of postjunctional supersensitivity.

B. Characteristics of Postjunctional Supersensitivity

1. Slow-Time Course of Development

The relationship of time to the development of postjunctional supersensitivity has been most extensively studies in the nictitating membrane of the cat. FLEMING and TRENDELENBURG (1961) found that 24 hours after a single dose of reserpine, in spite of apparently maximal depletion of endogenous noradrenaline, the nictitating membrane of the spinal cat was not supersensitive. However, a relatively small dose of reserpine (0.1 mg/kg) did produce supersensitivity after 7–14 days of daily administration. They concluded that time is essential for the development of supersensitivity and that there is no direct correlation between the depletion of noradrenaline and the increase in sensitivity. Unfortunately, this has often been misinterpreted to mean that the authors did not consider depletion to be a causative factor in the development of supersensitivity. In fact, as discussed in Section II, depletion probably is the critical effect of reserpine in inducing supersensitivity but there is a time lag between the cause and the effect.

Depletion was indirectly estimated by FLEMING and TRENDELENBURG (1961) from observations of the responses to nerve stimulation. Subsequently, direct chemical determinations (TRENDELENBURG and WEINER, 1962) established the validity of these estimates.

In 1963 (a) FLEMING presented results, based on various doses of reserpine and schedules of pretreatment, indicating that supersensitivity was present after 7 days of treatment but was considerably greater after 14 days. There was a trend toward a slightly greater sensitivity after 28 days but the difference between 14 and 28 days was not significant. On the basis of geometric means one can compute shifts of the following magnitude from FLEMING's (1963 a) data: 7 days, 4.2-fold; 14 days, 11-fold and 28 days, 14-fold. Fourteen days also appeared to be the optimal time for postjunctional supersensitivity to reach its maximum after decentralization or denervation (TRENDELENBURG, 1963 b).

After the discovery that the phenomenon of tone causes an underestimation of the shifts of the dose-response curves in nictitating membrane (see Section II), LANGER et al. (1967) reinvestigated the time course of the development of supersensitivity in the decentralized nictitating membrane. By using pithed (both brain and spinal cord destroyed) rather than spinal (brain destroyed, spinal cord intact) cats LANGER et al. were able to virtually abolish tone. Under these conditions there was a small but significant increase in sensitivity from the 14th to 28th day after decentralization or denervation. There would have been a considerable amount of tone in the decentralized and denervated nictitating membranes of TRENDELENBURG's (1963 b) experiments. It is also probable that there was tone in FLEMING's (1963 a) experiments. Depletion of catecholamine stores by reserpine would have reduced the tone in his experiments, but not as much as pithing the spinal cord (LANGER, 1966; LANGER et al., 1967). Some residual tone therefore may have obscured a small increase in sensitivity between the 14th and 28th days. It is clear, therefore, that the full development of postjunctional supersensitivity in the nictitating membrane requires 14 to 28 days to reach its maximum, regardless of whether it is induced by decentralization, denervation or chronic administration of reserpine.

A comparable period for the development of postjunctional supersensitivity in the iris sphincter of the cat has been found by GILBERT and FLEMING (unpublished results). Supersensitivity was maximal 3 weeks after removal of the ciliary ganglion.

The time required for maximum postjunctional supersensitivity to develop does vary with the tissue and perhaps with the species. FLEMING and TRENDELENBURG (1961) noted that supersensitivity of the cardiovascular system of the cat developed much more quickly than supersensitivity of the nictitating membrane. The optimum period of pretreatment with reserpine for development of supersensitivity of the heart rate and blood pressure responses of the cat was approximately three days. Consistent with these findings are the results of WESTFALL and FLEMING (1968b), in which there was supersensitivity to the chronotropic effects of noradrenaline or calcium in the dog heart lung preparation after three days of reserpine pretreatment but not after one day. Three days may not have been optimal, however, since longer periods of pretreatment were not tested. In the guinea pig heart the optimum period of pretreatment with reserpine is seven days (WESTFALL and FLEMING 1968a). Supersensitivity in vascular smooth muscle also develops quickly. Significant supersensitivity is present in rabbit aortic strips 24 hours after a single dose reserpine and is maximum after three daily doses (HUDGINS and FLEMING, 1966).

The only other tissue in which a careful study of the time course has been done is the vas deferens. WESTFALL (1970a) demonstrated that neither reserpine nor decentralization caused supersensitivity of the guinea pig vas deferens in 24 hours. Chronic administration of reserpine or decentralization caused significant shifts of the dose-response curves of various agonists five days after the treatment was begun. Ten days after decentralization, the shift of the norepinephrine curve was no greater than it was five days after decentralization. Hence, supersensitivity had developed fully within five days. KASUYA et al. (1969) found nonspecific supersensitivity to be maximal 4 days after denervation in the rat vas deferens.

Postjunctional supersensitivity develops in the guinea pig ileum after chronic administration of chlorisondamine, a ganglionic blocking agent (FLEMING, 1968). There is no supersensitivity 24 hours after treatment is begun, but the phenomenon is well developed after 5 to 7 days of treatment. Longer periods were not examined.

Thus, with few exceptions, postjunctional supersensitivity does not appear during the first 24 hours after the initiation of the sensitizing procedure and, depending on the tissue involved, will not reach its maximum until the sensitizing procedure has been maintained for three days to several weeks. Some of the studies in which supersensitivity has been anticipated but was not observed can be traced to the time factor. For example, WAUD and KRAYER (1960) found no supersensitivity in the dog heart-lung preparation 24 hours after a dose of reserpine which produces good depletion. Subsequently, one day was found to be an inadequate duration of treatment to produce supersensitivity in the dog heart (WESTFALL and FLEMING, 1968b).

2. The Nonspecific Nature

Some appreciation of the nonspecific nature of postjunctional supersensitivity may be obtained by inspecting Table 1. An explanation of the selection of data for Table 1 is in order. Only treatments which produced pronounced postjunctional supersensitivity have been included. In most instances these treatments have been proven to be optimal or nearly so. Any data in which prejunctional supersensitivity could be a complicating factor have been excluded. Thus no data are included which combine adrenergically denervated tissue and agonists such as noradrenaline, which are susceptible to uptake into adrenergic neurons. Furthermore, only data were used which could be presented in terms of ratios of geometric equieffective doses. This was possible whenever (1) the original data had been presented in that form (2) the data had been presented in a form readily and accurately converted to such or (3) the authors of this review had access to the original raw data.

Before discussing the details of individual tissues listed in the table, a few important generalizations concerning the nonspecificity of the supersensitivity will be stressed. First, in nine of the preparations the sensitivity was increased to two or more agonists which do not act via common receptors. In eight of these, supersensitivity to one or more ions (K^+, Ca^{++} or Ba^{++}) has been found. The evidence indicates that these ions are not acting through traditional drug receptors. Those examples in Table 1 in which no indication of nonspecificity is given have been taken from studies in which the question of supersensitivity to substances other than the natural transmitter or its congeners was not or could not be investigated. Few authors have considered the important question of specificity in supersensitivity.

In order to conclude that an example of supersensitivity is nonspecific one must take steps to eliminate the possibility of different agonists acting via the same receptors or of an agonist acting indirectly by releasing an active endogenous substance. This can be achieved by the use of specific blocking agents and/or procedures which eliminate a possible interfering endogenous substance.

Space does not allow a detailed analysis of these procedures in each tissue. For this purpose the reader is referred to the original references. For emphasis one example can be given from the work of HUDGINS and FLEMING (1966) on rabbit aortic strips. Chronic treatment with reserpine increased the sensitivity to noradrenaline, acetylcholine and potassium, but markedly decreased the sensitivity to tyramine which acts by the release of endogenous noradrenaline. If acetylcholine and potassium were acting by the release of noradrenaline, their dose-response curves should have been shifted to the right rather than the left. Furthermore, phentolamine, an adrenergic blocking agent, antagonized the effects of noradrenaline and tyramine but not those of acetylcholine or potassium. It is notable in this context that phentolamine is extremely effective in antagonizing the effects of endogenously released noradrenaline in the rabbit aortic strip (HUDGINS and FLEMING, 1966; URQUILLA et al., 1970). Thus neither acetylcholine nor potassium can act by the release of noradrenaline, nor can they act directly on the adrenergic receptors.

The second important generalization to be made from the table is that there is not a consistently greater supersensitivity to the natural transmitter and/or its chemical derivatives in comparison to other unrelated agonists. This matter will be discussed in detail below for each of the tissues.

3. Characteristics of Postjunctional Supersensitivity in Individual Tissues

a) Nictitating Membrane

In studies of supersensitivity more work has been done with the nictitating membrane of the cat than any other organ. The data in Table 1 reflects a small proportion of the data in the literature. Much of the earlier work has been reviewed very effectively by TRENDELENBURG (1963 a) and is therefore omitted. Furthermore, to make valid quantitative comparisons the only data included are those obtained in experiments in which tone (LANGER, 1966) had been reduced and in which the time allowed for postjunctional supersensitivity was approximately optimal (14 to 28 days).

The data in Table 1 illustrate that the increase in sensitivity to (−)-noradrenaline (14- to 25-fold) is only 2–3 times greater than that to nonadrenergic agents such as acetylcholine and barium (7- to 8-fold). Such a difference is small and cannot be used to argue in favor of an involvement of adrenergic receptors in the mechanism of postjunctional supersensitivity because the shifts in sensitivity to other adrenergic agonists were smaller. SEIDEHAMEL et al. (1966) reported a 2.9-fold increase in sensitivity to (+)-noradrenaline

Table 1. Postjunctional supersensitivity in a variety of tissues

Species	Tissue	In vitro[a]	Treatment Type	Days	Agonist	Shift[b]	Reference
Cat	Nictitating membrane	—	Dec	28	1-NA	16	LANGER et al. (1967)[c]
		—	R	28	1-NA	14	FLEMING (1963a)
		—	R	14	1-NA	16.2	SEIDEHAMEL et al. (1966)
		—	R+Dec	14	1-NA	25	FLEMING (1963a)
		—	Dec	28	d-NA	2	LANGER et al. (1967)[c]
		—	R	14	d-NA	2.9	SEIDEHAMEL et al. (1966)
		—	Dec	28	1-Ad	16.7	LANGER et al. (1967)[c]
		—	Dec	28	Methox.	8.9	TRENDELENBURG et al. (1970)[c]
		—	R+Dec	14	ACh	8.2	FLEMING (1963a)
		—	Dec	28	ACh	8	LANGER et al. (1967)[c]
		—	Dec	14–21	Ba++	7.1	MORRISON and FLEMING (1971)[c]
		—	Den	14–21	Ba++	7.4	MORRISON and FLEMING (1971)[c]
Guinea pig	Vas deferens	yes	Dec	5	1-NA	29	WESTFALL (1970a)
		yes	R	5	1-Na	29	WESTFALL (1970a)
		yes	Dec	7	1-NA	2.5	WESTFALL et al. (1972)
		yes	Dec	5	MeFur	6.2	WESTFALL (1970a)
		yes	R	5	MeFur	6.5	WESTFALL (1970a)
		yes	Dec	5	Hist	14	WESTFALL (1970a)
		yes	R	5	Hist	16.6	WESTFALL (1970a)
		yes	Dec	7	Hist	2	WESTFALL et al. (1972)
		yes	Den	7	Hist	3.7	WESTFALL et al. (1972)
		yes	Dec	5	K+	1.5	WESTFALL (1970a)
		yes	R	5	K+	1.4	WESTFALL (1970a)
		yes	Den	7	K+	2.3	WESTFALL et al. (1972)
Rat	Vas deferens	yes	Den	7	ACh	30	LEE, WESTFALL and FLEMING (unpublished)
		yes	Den	7	ACh	4.5	FOLEY, McPHILLIPS and WESTFALL (unpublished)
		yes	Den	7	K+	4.0	WESTFALL (unpublished)
Guinea pig	Ileum	yes	Gang. bl.	5–7	ACh	45[d]	FLEMING (1968)
		yes	Gang. bl.	5–7	5-HT[c]	3.6	FLEMING (1968)
		yes	Gang. bl.	5–7	Hist	3.8	FLEMING (1968)
		yes	Gang. bl.	5–7	K+	7.8	FLEMING (1968)

Animal	Tissue	[a]	Treatment	No.	Agonist	Ratio[b]	Reference
Cat	Iris sphincter	—	Den	35+	Pilo	8	Brto and Dawson (1970)
		—	Den	35+	Carb	20	Brto and Dawson (1970)
		yes	Den	21	ACh	8.2	Gilbert and Fleming (unpublished)
		yes	Den	21	Hist	6.7	Gilbert and Fleming (unpublished)
Pig	Iris sphincter	yes	Den	14	ACh	8.7	Schaeppi (1966)
Rabbit	Aortic strip	yes	R	3	1-NA	17	Hudgins and Fleming (1966)
		yes	R	3	Phenyleph.	2.9	Taylor and Green (1971)
		yes	R	3	ACh	3	Hudgins and Fleming (1966)
		yes	R	3	ACh	4.3	Taylor and Green (1971)
		yes	R	3	K+	1.7	Hudgins and Fleming (1966)
		yes	R	3	K+	2.2	Taylor and Green (1971)
		yes	6-OHDA	1	K+	1.9	Shibita et al. (1972)
		yes	R	3	Ba++	2	Taylor and Green (1971)
		yes	6-OHDA	1	Ba++	1.7	Shibita et al. (1972)
Rabbit	Ear artery	yes	Dec	14	1-NA	2.1	De la Lande and Fleming (unpublished)
		yes	Dec	14	K+	1.3	De la Lande and Fleming (unpublished)
Dog	Mesenteric artery	yes	R	≅360	1-NA	2.1	Clarke et al. (1970)
Rat	Hind limb	—	R	5	1-NA	1.8	Brody et al. (1964)
		—	R	5	1-Ad	2.1	Brody et al. (1964)
Dog	Heart (HLP)	semi	R	3	1-NA	2	Westfall and Fleming (1968b)
		semi	R	3	Ca++	1.8	Westfall and Fleming (1968b)
Guinea pig	Heart	—	R	7	1-NA	7.5	Westfall and Fleming (1968a)
		yes	R	7	1-NA	2	Westfall and Fleming (1968a)
		yes	R	7	Ca++	3	Westfall and Fleming (1968a)
Rat	Heart	—	Gang. bl.	6	Carb	2.8	Corey and McPhillips (1972)
		—	Gang. bl.	6	MeCh	2.1	Corey and McPhillips (1972)
Cat	Sweat glands	—	Den	7	ACh	60	Reas and Trendelenburg (1967)
		—	Den	7	Pilo	25	Reas and Trendelenburg (1967)
		—	Gang. bl.	4	ACh	20	Reas and Trendelenburg (1967)
		—	Gang. bl.	4	Pilo	10	Reas and Trendelenburg (1967)

[a] Indicates that the sensitivity was determined in isolated tissues after the treatment period of the animal had been completed.
[b] Ratio of geometric mean of equieffective doses in control and supersensitive tissue or its equivalent.
[c] Experiments done in pithed cats to cause maximum reduction in tone of the supersensitive nictitating membranes.
[d] May be an overestimation. — See text.
[e] 5-HT does not act directly on ileum. — See text.

Abbreviations: R = reserpine; Dec = decentralization; Den = denervation; ACh = acetylcholine; Gang. bl. = ganglionic blockade; 6-OHDA = 6-hydroxydopamine; NA = noradrenaline; Ad = adrenaline; Methox. = methoxamine; MeFur = methylfurmethide; Hist. = histamine; 5-HT = 5-hydroxytryptamine; Carb. = carbachol; Phenyleph = phenylephrine; MeCh = methacholine; Pilo = pilocarpine.

after 14 days of treatment with reserpine. LANGER et al. (1967) found a very similar shift of the (+)-noradrenaline curve, 2-fold, 28 days after decentralization. Decentralization produced an 8.9-fold increase in sensitivity to methoxamine (TRENDELENBURG et al., 1970). The reason for these moderate differences in the shift of the dose-response curve among different agonists is not known. However, LANGER and TRENDELENBURG (1969) have presented a plausible explanation for the stereo-selectivity of postjunctional supersensitivity to isomers of noradrenaline. Because of the low potency of (+)-noradrenaline, its uptake by adrenergic neurons is relatively saturated in the range of its normal dose-response curve. An increase in the sensitivity of the effector cells would shift the dose-response curve to the left and into a range of doses in which uptake is more efficient. Thus, the shift to the left due to increased effector sensitivity is partly countered by increased uptake of the amine (see also Section II).

Another appealing theory which may explain differences in the magnitude of the increase in sensitivity to a wider variety of agonists has been proposed by WESTFALL (1970a) and others (see Section III. C. 2). This hypothesis is based upon differences among agonists in their mechanism of inducing excitation of the effector cells.

TRENDELENBURG and WEINER (1962) demonstrated that chronic treatment with either reserpine or the adrenergic neuron-blocking substance, 2:6-xylyl ether bromide (TM-10), produced supersensitivity of the nictitating membrane to noradrenaline or acetylcholine of a magnitude comparable to that produced by decentralization. A somewhat smaller increase in sensitivity was produced by chronic treatment with chlorisondamine, a finding attributable to incomplete ganglionic blockade (TRENDELENBURG and WEINER, 1962).

A more precise estimate of the shifts of dose-response curves produced by different procedures can be obtained from the data in Table 1. For example, 28 days of decentralization and 28 days of reserpine administration increased the sensitivity to (−)-noradrenaline 16- and 14-fold respectively. Furthermore, the effects of chronic reserpine administration and decentralization are not additive. The two procedures together increased the sensitivity to noradrenaline only 25-fold. If they were additive one would expect an increase insensitivity of over 200-fold (a figure which is derived by adding the individual *log* shifts induced by each procedure separately). It should also be noted that denervation or decentralization caused equal increases in sensitivity to barium.

Thus the magnitude of postjunctional supersensitivity to any one agonist is approximately the same regardless of whether it is induced by denervation, decentralization, chronic depletion of transmitter, chronic neuron blockade or chronic ganglionic blockade. It is clear that these procedures are producing supersensitivity through a common mechanism (see EMMELIN, 1961; TRENDELENBURG and WEINER, 1962; Section III. A). These results also suggest modifications for the "Law of Denervation" (CANNON and ROSENBLUETH, 1949). (1) Supersensitivity is produced by pharmacological as well as surgical interruption of innervation (EMMELIN, 1961). (2) The magnitude of super-sensitivity does not necessarily decrease as the interruption of innervation is made at greater distances from the effector, *if* one eliminates the complicating factor of prejunctional supersensitivity.

Supersensitivity of the denervated nictitating membrane to 5-hydroxy-tryptamine has been reported by SCHAEPPI (1963). However, interpretation of the data is complex since the drug is taken up by adrenergic neurons and acts in part indirectly by stimulating the superior carvical ganglion (TREN-DELENBURG, 1956). Thus, denervation would decrease responses to 5-HT by removing its indirect component of action and simultaneously increase the sensitivity to it by both pre- and postjunctional supersensitivity. In a recent study, PLUCHINO (1972) has eliminated indirect actions of 5-HT on the nicti-tating membrane by a combination of adrenalectomy and short-term pre-treatment with reserpine. Under these conditions, PLUCHINO was able to clearly demonstrate supersensitivity of the chronically decentralized nicti-tating membrane to the direct action of 5-HT.

In a preliminary communication (FLEMING, 1963b) denervation was re-ported to produce supersensitivity to potassium. However, subsequent in-vestigation of the problem indicated that a considerable portion of the response to intraarterially administered potassium was due to release of catecholamines, probably from sources outside the nictitating membrane. Thus, the question of whether or not the sensitivity of the nictitating membrane increases to the direct action of potassium remains unsettled.

Isoproterenol has a relaxant effect on the partially contracted nictitating membrane and this action is mediated by beta-adrenergic receptors (SMITH, 1963). Supersensitivity to this inhibitory effect is produced neither by de-nervation (SMTIH, 1963) nor by decentralization (PLUCHINO and TRENDELEN-BURG, 1968). On the other hand, the sensitivity to the alpha-stimulatory effect of isoproterenol in the presence of beta blockade is increased by decentraliza-tion (PLUCHINO and TRENDELENBURG, 1968).

It can be concluded that postjunctional supersensitivity in the nictitating membrane is associated with increased sensitivity to the direct stimulant actions of sympathomimetics, acetylcholine, 5-HT and barium. The evidence is thus far inconclusive regarding potassium. On the other hand, the sensi-tivity is *not* increased to the inhibitory actions of sympathomimetics.

b) Vas Deferens

As indicated in Table 1, postjunctional supersensitivity has been demon-strated in the guinea pig vas deferens after chronic denervation, decentralization or reserpine administration and in the rat vas deferens after denervation. WESTFALL (1970a) found that, similar to the results in the nictitating membrane, the increase in sensitivity to any one agonist was very similar regardless of whether it was induced by decentralization or by administration of reserpine. WESTFALL (1970a) also found that the shift was greatest for noradrenaline (29-fold) and less for the other agonists (histamine, 14-fold; methylfurmethide, 6-fold and potassium, 1.5-fold). However, in a subsequent study (WESTFALL et al., 1972) no such differences among the shifts of dose-response curves appeared. Methylfurmethide was not included in this latter study but the increases were quite similar for noradrenaline (2.5-fold), histamine (2–3.7-fold) and potassium (2.3-fold). The reasons for the quantitative differences between these two studies are not known. They were carried out by the same investigator but in different laboratories (Oxford and West Virginia). Thus the housing environments of the animals differed and there may have been some genetic difference between the groups of animals.

Denervation of the rat vas deferens has also been found to increase the sensitivity nonspecifically (Table 1). It is interesting that LEE, WESTFALL and FLEMING (unpublished results) found a 30-fold increase in the sensitivity of the denervated rat vas deferens to acetylcholine, which is approximately the increase reported by KASUYA et al. (1969), as estimated from the dose-response curves of the latter authors. In contrast, FOLEY, McPHILLIPS and WESTFALL (unpublished results) found only a 4.5-fold increase in the sensitivity to carbachol was produced by denervation in the rat vas deferens. This discrepancy is readily explained, since denervation caused a 75 % decrease in the cholinesterase activity of the denervated vas deferens of both rat and guinea pig (FOLEY, McPHILLIPS and WESTFALL, unpublished results). This loss of cholinesterase activity is presumably due to degeneration of postganglionic cholinergic neurons which have been shown to be present in the vas deferens (BURNSTOCK, 1970; INOMATA and SUZUKI, 1971). The 4.5-fold shift in the carbachol curve thus represents postjunctional supersensitivity and the greater shift in the acetylcholine curve represents the additive effect of the decreased degradation of acetylcholine (i.e. a prejunctional supersensitivity). Changes in cholinesterase did not play any role in the results shown in the table for the guinea pig vas deferens, since none of the agonists used are substrates for cholinesterase. It is also noteworthy that decentralization does not change the cholinesterase activity of the guinea pig vas deferens (FOLEY, McPHILLIPS and WESTFALL, unpublished results).

There is controversy over the question of whether or not postjunctional supersensitivity does occur in the vas deferens. As described above, WESTFALL (1970a) and WESTFALL et al. (1972) have provided extensive data supporting the existence of postjunctional supersensitivity in the guinea pig vas deferens produced by denervation, decentralization or chronic administration of reserpine. Several studies have presented evidence of postjunctional supersensitivity in the denervated vas deferens of the rat (KASUYA et al., 1969; BIRMINGHAM et al., 1970; unpublished results from our laboratory).

In contrast, other groups have failed to find evidence of postjunctional supersensitivity in the vas deferens. OZAWA and SUGAWARA (1970) attempted, unsuccessfully, to find evidence for postsynaptic supersensitivity in the guinea pig vas deferens after a variety of chronic procedures including denervation, decentralization and the administration of reserpine, hexamethonium and tetrodotoxin. However, one must question whether the procedures were adequate to stop neuromuscular transmission and/or maintain that interruption over a period of several days. Each of these will be considered briefly.

OZAWA and SUGAWARA (1970) observed a shift of less than 5-fold in the noradrenaline dose-response curve after denervation. This is quite small even if the denervation had produced only prejunctional supersensitivity. WAKADE and KIRPEKAR (1971) obtained a shift of approximately 30-fold in the denervated guinea pig vas. OZAWA and SUGAWARA did not measure tissue catecholamines to determine the completeness of denervation. They did report that 10^{-6} M cocaine caused no further shift of the noradrenaline dose-response curve in the denervated tissues. However, 10^{-6} M cocaine is a suboptimal concentration. Indeed, it produced only a 2-fold shift in control tissues (OZAWA and SUGAWARA, 1970). WESTFALL et al. (1972) observed a 13-fold increase in sensitivity to noradrenaline with 10^{-5} M cocaine.

WESTFALL et al. (1972) found that, in a sizeable number of operations, decentralization of the guinea pig vas was not completely successful because of branching of the hypogastric nerves. To be certain that only completely decentralized preparations were used in their experiments, WESTFALL et al. electrically stimulated the nerve on the central side of the cut section just before the vas deferens was removed from the animal on the day of the experiment. If the vas deferens responded at all to nerve stimulation it was considered to be partially innervated and discarded. OZAWA and SUGAWARA (1970) gave no indication that such precautions were taken in their work.

OZAWA and SUGAWARA (1970) administered reserpine on two dose-schedules, 3 mg/kg/day for two days and 0.1 mg/kg/day for 7 or 14 days. The former is probably too short a period to produce supersensitivity (WESTFALL, 1970a) and the latter is too small a dose of reserpine. McCLURE and WESTFALL (unpublished results)observed noradrenaline concentrations of 8.19 ± 0.66 (N = 16) and 1.80 ± 0.26 μg/g wet weight (N = 16) respectively in control vasa deferentia and vasa obtained from animals pretreated with seven daily injections of reserpine, 0.1 mg/kg. This decrease of 78 % probably would not interrupt transmission. LEE (1967) found that quite good responses of the heart and nictitating membrane of the cat to nerve stimulation were obtained after 90 % depletion of their catecholamine stores.

Finally, OZAWA and SUGAWARA (1970) administered hexamethonium twice daily for seven days and tetrodotoxin in various daily doses for up to 28 days. Hexamethonium has a very short duration of action. Twice-daily injections could not be expected to depress ganglionic transmission throughout the day. If the tetrodotoxin were given in adequate doses to interrupt nerve transmission, one would expect the animals to die from interruption of the respiratory nerves. Thus one must have reservations about the adequacy of the treatments used by OZAWA and SUGAWARA (1970).

In recent preliminary communications there have been reports that decentralization does not produce supersensitivity in the vas deferens of the guinea pig (ROMANO et al., 1971) or the rat (TSAI and PENN, 1972). The reasons for these negative findings are unknown. However, there are enough positive results to convince the authors of this review that postjunctional supersensitivity does indeed occur in the vas deferens of the guinea pig and rat.

Two further points of interest should be made regarding the effects of chronic denervation of the vas deferens. KASUYA et al. (1969) noted that denervation of the rat vas deferens caused an increase in the maximum response to noradrenaline and acetylcholine but not to potassium. DE MORAES et al. (1970) reported an increased maximum response to noradrenaline in the denervated vas deferens of the guinea pig. In our laboratory this same phenomenon has been observed in both the guinea pig (WESTFALL et al., 1972) and the rat (LEE, WESTFALL and FLEMING, unpublished results). Furthermore, the latter two groups found that denervation prolonged the response to noradrenaline, acetylcholine and histamine, and accentuated the plateau phase of the response to potassium. Cocaine prolonged the response to noradrenaline only. Decentralization, at least in the guinea pig (WESTFALL et al., 1972) had no effect on maximum responses, duration of responses or the plateau phase of the response to potassium. The effects on maximum responses, duration of response and the plateau phase can only be interpreted as an indication that denervation causes some change in the smooth muscle cells over and above the postjunctional supersensitivity caused by both denervation and decentralization. The authors are unaware of any parallels to this in other tissues with the possible exception of the rabbit aorta in which chronically administered reserpine increases the maximum response specifically to acetylcholine (see Section III. C and TAYLOR and GREEN, 1971).

c) Ileum

Chronic treatment with a ganglionic blocking agent (chlorisondiamine, 1.0 mg/kg, every 8 hours for 5–7 days) produces nonspecific supersensitivity of the guinea pig ileum (FLEMING, 1968). The increase in sensitivity to acetylcholine was much greater than that to histamine or potassium. However the

45-fold increase in sensitivity may be an overestimation. This conclusion is based on the following observations: (1) The increase in sensitivity to 5-HT is only 3.6-fold. In the guinea pig ileum 5-HT acts entirely by the release of acetylcholine (DAY and VANE, 1963; BROWNLEE and JOHNSON, 1963, 1965). (2) The dose-response curve for acetylcholine was not shifted in a parallel fashion, in contrast to that for the other agonists. The lower part of the curve was shifted much more than the upper portion. (3) Accurate measurement of the responses of the supersensitive tissues to doses of acetylcholine in the lower dose ranges was difficult because they consisted primarily of a series of spikes rather than tonic contractions. This was in contrast to the tonic responses of control tissues to acetylcholine and of both tissues to the other agonists.

Chronic treatment of guinea pigs with reserpine (1.0 mg/kg/day for 5-7 days) decreased the noradrenaline content of the ileum by 95 % but did not increase the sensitivity to acetylcholine, histamine or potassium (GREEN et al., 1968). It would thus appear that the nonspecific supersensitivity in this tissue required the chronic interruption of the normal stimulatory (cholinergic) innervation, not of the inhibitory (adrenergic) innervation. It may be noted that GREEN et al. (1968) found that the chronic administration of reserpine increased the acetylcholine content of the ileum and did increase the sensitivity specifically to agents which act by releasing acetylcholine.

Although chronic reserpine administration does not affect the sensitivity of the guinea pig ileum to directly acting stimulant agonists, it does produce supersensitivity of the isolated rabbit ileum to the inhibitory effects of directly acting sympathomimetics (SCHMIDT and FLEMING, 1963). Although this is presumed to be an expression of postjunctional supersensitivity, no increase in sensitivity to nonsympathomimetic inhibitory compounds, such as papaverine, has been found in this tissue (SCHMIDT and FLEMING, 1964).

d) Iris

Removal of the ciliary ganglion produces postjunctional supersensitivity in the iris sphincter of the cat (Table 1) measured in vivo (BITO and DAWSON, 1970) and in vitro (GILBERT and FLEMING, unpublished results). The increased sensitivity to acetylcholine observed in vitro was of the same magnitude as the increased sensitivity to pilocarpine and slightly less than half that to carbachol in vivo. There was an increased sensitivity to histamine, which was similar to that to acetylcholine. A similar increase in sensitivity to acetylcholine has been seen in the isolated iris of the pig (SCHAEPPI, 1966). On the other hand, LANGHAM and FRASER (1966) could not detect supersensitivity to noradrenaline in the intact iris dilator muscle of the rabbit up to 21 days after sympathetic decentralization. Supersensitivity of the pupils to methacholine has been reported in patients with familial disautonomia (SMITH and DANCIS, 1963) or degeneration of the autonomic nervous system (FREWIN et al., 1968).

e) Vascular Smooth Muscle

Rauwolfia alkaloids and other catecholamine-depleting substances have been shown to enhance the increase in the blood pressure produced by directly acting sympathomimetics. None of the studies considered the sensitivity of the blood pressure to non-sympathomimetic agonists. For example, ORLANS et al. (1960) reported that syrosingopine, administered to dogs in a dose of 0.5 mg/kg 24 hours before the experiment, produced an approximately 3-fold shift of the dose-response curve of noradrenaline pressor effects to the left. MAXWELL et al. (1960) found that, 48 hours after a 15 mg/kg dose of guanethidine, the responses of the blood pressure of dogs to single doses of noradrenaline, adrenaline, cobefrin and epinine were significantly greater than the responses of control dogs. Although guanethidine has a cocaine-like effect as well as an amine-depleting action, the former would not be expected to last for 48 hours. Thus the supersensitivity was probably of the postjunctional type.

FLEMING and TRENDELENBURG (1961) showed that the dose-response curve of the pressor response of the spinal cat to noradrenaline was shifted to the left approximately 3-fold by 1–3 days of various daily doses of reserpine. VAN ZWIETEN et al. (1965) found enhanced pressor responses to noradrenaline in pithed rats pretreated with reserpine, 3 mg/kg, 42 and 18 hours before the experiment. It is important to note that the cats used by FLEMING and TRENDELENBURG were spinal (i.e. the spinal column cut in the cervical region and the brain pithed) and the rats used by VAN ZWIETEN et al. were pithed. These methods eliminate much of the tone exerted by the autonomic nervous system on the cardiovascular system. Thus, under these conditions, there is little difference in the resting blood pressure between reserpine-pretreated animals and controls, and responses are not complicated by cardiovascular reflexes.

There is little information in the literature regarding the existence of postjunctional supersensitivity in the vascular system of man. ABBOUD and ECKSTEIN (1964a) tested the cardiovascular responses of seven subjects to infusions of noradrenaline before, at the end of, and four weeks after completion of a 14-day treatment with reserpine, 1.0 mg per day. They reported no significant effect of reserpine on the systematic arterial pressor responses or forearm blood flow to noradrenaline even though decreased responses to tyramine indicated that some depletion of endogenous noradrenaline had occurred. However, there are two problems with their presentation of the data. First, they did not use a paired-sample statistical analysis, even though each stubjec served as his own control. Second, the "control" responses were the averages of the pre-reserpine and 4-week post-reserpine determinations. Inspection of the data suggests that responses were enhanced by reserpine but that the enhanced responses tended to remain even 4 weeks after the treatment was discontinued. Since recovery of catecholamine stores in rats and rabbits takes 4 to 7 weeks after a single large dose of reserpine (DAHLSTRÖM and HÄGGEN-DAL, 1966) one must recognize the possibility that normal nerve transmission (and thus normal sensitivity) is not completely restored in man 4 weeks after the end of chronic treatment.

Fortunately, ABBOUD and ECKSTEIN (1964a) provided individual data on pressor responses for each subject with one rate of infusion of noradrenaline (0.15 µg/kg/min) in Table 1 of their paper. We have reanalyzed these data. Paired-sample analysis of the responses on the 14th day of reserpine in comparison with those 4 weeks after the end of treatment indicate that they were not different from each other. However, each of them was significantly greater than the responses obtained before reserpine (p values below 0.05). Therefore, the results tend to indicate that chronic administration of reserpine *does* increase the sensitivity of the human cardiovascular system to noradrenaline.

The meaning of enhanced pressor responses is difficult to interpret because blood pressure responses are complex, integtrated responses of heart rate, cardiac stroke volume and changes in resistance in separate vascular beds. Recognizing this problem, WITHRINGTON and ZAIMIS (1967) investigated blood pressure, chronotropic, inotropic and hind limb blood-flow responses of control cats and cats pretreated with reserpine, 10 µg/kg/day for 165 to 185 days. Reflexes and differences in baseline were reduced by means of acute ganglionic blockade. These investigators found the cardiac and pressor responses to noradrenaline and adrenaline were enhanced in the treated animals, whereas the responses of the hind limb were unaltered. They concluded that the enhanced responses of the blood pressure were secondary to cardiac supersensitivity and that the sensitivity of the vascular system per se was not changed.

However, other workers have reported supersensitivity, presumably of the postjunctional type, in specific vascular beds. ABBOUD and ECKSTEIN (1964b) reported an enhanced response of the brachial artery pressure to noradrenaline in dogs pretreated with reserpine, 0.25 mg/kg, 48 and 24 hours before the experiment. BRODY and DIXON (1964) observed supersensitivity to the pressor effects of noradrenaline and adrenaline in the perfused hindquarters of the rat pretreated with reserpine, 1 mg/kg/day for 5 days (Table 1). The sensitivity appeared to be specific since the responses to single doses of angiotensin, vasopressin and barium were no greater than control. It is possible, of course, that dose-response data might have uncovered an effect. In contrast, SAKURAI and HASHIMOTO (1965), in experiments on the perfused ear of the rabbit, observed a 2-fold increase in sensitivity to noradrenaline and a 5-fold increase in sensitivity to angiotensin 24 hours after pretreatment with reserpine, 10 mg/kg.

In isolated perfused mesenteric blood vessels removed from dogs treated with reserpine, 0.137 mg *per os* twice daily for a year, CLARKE et al. (1970) found a 2.1-fold shift to the left of the noradrenaline dose-response curve (Table 1). They did not study the effects of nonsympathomimetic agonists. DE LA LANDE and FLEMING (unpublished results) observed significant increases in the sensitivity of perfused ear arteries from rabbits which had been chronically decentralized. The sensitivity to both noradrenaline and potassium was increased (Table 1). These workers also noted increased responses to single doses of angiotensin in the decentralized arteries.

Several different groups of investigators have reported postjunctional supersensitivity in isolated strips of rabbit aorta. The first were BURN and RAND (1958) who found that pretreatment with reserpine induced supersensitivity to noradrenaline. Later work (Table 1) has established that the sensitivity is increased to several nonadrenergic agonists as well as noradrenalin by either chronic treatment with reserpine (HUDGINS and FLEMING, 1966; TAYLOR and GREEN, 1971; GARRETT and CARRIER, 1971) or destruction of adrenergic nerves by pretreatment with 6-hydroxydopamine (MALING et al., 1971; SHIBATA et al., 1972). It should be noted that the supersensitivity may be less than maximal in some of these studies, since 3 days of treatment is optimal for this tissue (HUDGINS and FLEMING, 1966). The increased sensitivity to adrenergic amines produced by 6-hydroxydopamine will not be considered here since it is due in part to prejunctional supersensitivity. An increased sensitivity was demonstrated to noradrenaline (HUDGINS and FLEMING, 1966); phenylephrine (TAYLOR and GREEN, 1971); acetylcholine (HUDGINS and FLEMING, 1966; TAYLOR and GREEN, 1971); potassium (HUDGINS and FLEMING, 1966; TAYLOR and GREEN, 1971; MALING et al., 1971; GARRETT and CARRIER, 1971; SHIBATA et al., 1972); and barium (TAYLOR and GREEN, 1971; SHIBATA et al., 1972).

This supersensitivity of the aorta has been termed relatively nonspecific because there was no increase in sensitivity to histamine (HUDGINS and FLEMING, 1966; TAYLOR and GREEN, 1971), angiotensin (HUDGINS and FLEMING, 1966) or 5-hydroxytryptamine (HUDGINS and FLEMING, 1966). It is possible that angiotensin and 5-HT produce part of their effect on the aortic strip by releasing noradrenaline. If so, an increased sensitivity to any direct action which they possess may have been masked by the depletion of endogenous noradrenaline.

Although HUDGINS and FLEMING (1966) observed a relatively large shift of the noradrenaline dose-response curve, TAYLOR and GREEN (1971) found a shift of only 2.9 for another directly acting sympathomimetic, phenylephrine. The latter value is close to the shifts reported by all of the authors cited above for nonadrenergic agonists (1.7 to 4.3). It is also of interest that TAYLOR and GREEN (1971) observed a 16-fold shift of the acetylcholine curve in one comparison involving 4 controls and 4 experimental preparations but only 4.3-fold in a subsequent comparison of larger groups.

MALING et al. (1971) did not observe any supersensitivity of aortic strips of rabbits, guinea pigs or rats 20–24 hours after pretreatment with reserpine, 3 mg/kg. Although HUDGINS and FLEMING (1966) did find evidence of supersensitivity 1 day after reserpine administration, the latter authors established that 3 days of treatment was optimal. In the study of MALING et al. (1971) 6-hydroxydopamine produced supersensitivity in aortic strips from rabbits but not from guinea pigs or rats. The rat has almost no endogenous noradrenaline in the aorta. This fact suggests that the adrenergic nerves play little, if any, role in the normal function of the rat aorta. Thus, 6-hydroxydopamine would not be expected to have an effect. This explanation does not apply to the guinea pig, the aorta of which has more noradrenaline than the aorta of the rabbit (0.93 µg/g vs. 0.41 µg/g, MALING et al., 1971). However, the depletion of noradrenaline by 6-hydroxydopamine was somewhat less in the guinea pig aorta than in the rabbit aorta.

Large vessels have a peculiar innervation (BURNSTOCK, 1970). The adrenergic nerves are generally confined to the adventitial border and the distances between even the outermost smooth muscle cells and the nerve terminals are large (> 800 Å). This may be compensated for by intercellular contacts which allow excitation to spread from one muscle cell to another (BURNSTOCK, 1970). It is likely that variations in the density of nerve terminals (as indicated by the noradrenaline content), in the magnitude of the neuromuscular gaps and in the frequency of intermuscular electrical contacts all help to determine the magnitude of neuronal regulation of aortic smooth muscle in different species. If the adrenergic nerves have a very small influence on the muscle cells, little or no supersensitivity would be expected to result from interruption, of the neuromuscular transmission.

The very thorough study by TAYLOR and GREEN (1971) established that, in the isolated aorta of the rabbit, the supersensitivity induced by chronic administration of reserpine is associated with an increased maximum response to acetylcholine, but not to phenylephrine, potassium or barium (see Section III. C).

f) Cardiac Muscle

Depending on the species, supersensitivity to the chronotropic effects of drugs appears after 1 to 3 days of treatment with reserpine and reaches a maximum after 3 to 7 days of treatment (FLEMING and TRENDELENBURG, 1961; WESTFALL and FLEMING, 1968a, b). In the dog (WESTFALL and FLEMING, 1968b) and the guinea pig (WESTFALL and FLEMING, 1968a) the sensitivity to both noradrenaline and calcium was increased to equivalent degrees. An attempt to show supersensitivity to theophylline in the dog heart was unsuccessful due to the fact that theophylline was found to act, in part, via the release of endogenous noradrenaline (Westfall and Fleming, 1968b).

As indicated in Table 1, the magnitude of supersensitivity observed in the intact heart (pithed guinea pigs) was greater than that observed in isolated perfused guinea pig hearts. The significance of this will be discussed in Section III. D.

One study of supersensitivity of the heart to negative chronotropic agonists has been made. Corey and McPhillips (1972) treated rats with the ganglionic blocking agent, chlorisondamine, twice daily for 6 days. Subsequently the *in vivo* sensitivity of the hearts of the treated animals to carbachol and methacholine was higher than in controls.

The supersensitivity to chronotropic agonists induced by chronic treatment with reserpine is accompanied by an increased sensitivity to drug-induced arrythmias in the cat (Fleming, 1962) and guinea pig (Westfall and Fleming, 1968a). In the latter instance the arrhythmic responses to both noradrenaline and calcium were shown to be enhanced. It is interesting that Lown et al. (1961) found that administration of therapeutic doses of reserpine for three days enhanced the sensitivity of patients with atrial fibrillation to arrhythmias induced by acetylstrophantidin. Muelheims et al. (1965) reported supersensitivity to catecholamine-induced arrhythmias in normal subjects receiving guanethidine daily for three days. In the experiments by Muelheims and coworkers one cannot absolutely rule out contributions of blood pressure increases and prejunctional supersensitivity as well as postjunctional supersensitivity to the arrhythmias.

On the other hand, postjunctional supersensitivity to cardiac inotropic effects may not occur. Dempsey and Cooper (1968) demonstrated cat hearts by neural ablation. After a period of 6–23 days the hearts were removed, perfused and electrically driven. In comparison to control hearts the denervated hearts were supersensitive to the inotropic effects of noradrenaline but not to those of isoproterenol or calcium. The authors concluded that the hearts demonstrated prejunctional supersensitivity but not postjunctional. In contrast, McNeill and Schulze (1972) reported supersensitivity of perfused guinea pig hearts from animals given large doses of reserpine 48 and 24 hours previously. Unfortunately the latter studies were made in spontaneously beating hearts. Since the interval between beats affects contractility (Koch-Weser and Blinks, 1963), one cannot adequately judge the sensitivity of spontaneously beating hearts to inotropic drugs which also alter rate. In preliminary experiments, Taylor, Westfall and Fleming (unpublished results) have been unable to detect supersensitivity of electrically-paced perfused hearts of guinea pigs under conditions which produce supersensitivity to the chronotropic effects of noradrenaline and calcium in spontaneously beating hearts.

g) Exocrine Glands

The topic of nonspecific supersensitivity in salivary glands has been thoroughly reviewed by EMMELIN (1961, 1965). These reviews have been discussed in Section I. Suffice it to say at this point that supersensitivity of salivary glands develops after the same chronic procedures which produce the phenomenon in smooth and cardiac muscle. The supersensitivity develops gradually and is nonspecific. Two additional references should be added.

MÓZSIK et al. (1967) have demonstrated postjunctional supersensitivity in the parotid salivary glands of man. Subjects were treated with atropine (0.3 to 1.0 mg, 3 times per day for 2–4 weeks). During this period, tolerance of the glands developed to atropine along with supersensitivity to the secretory effects of both noradrenaline and acetylcholine. There was also a tendency toward greater sensitivity to histamine, although the difference was not statistically significant. Thus, not only does postjunctional supersensitivity occur in man but, by inducing supersensitivity drugs used therapeutically may cause "tolerance" to themselves.

Sweat glands also develop postjunctional supersensitivity (Table 1). REAS and TRENDELENBURG (1967) demonstrated supersensitivity of the sweat glands of the cat's paw produced by either denervation or chronic ganglionic blockade. The maximum effect required 7 days. REAS and TRENDELENBURG also demonstrated that chronic administration of pilocarpine after denervation prevented the supersensitivity from developing. The supersensitivity was clearly postjunctional because (1) pilocarpine and acetylcholine are not taken up into nerves and (2) possible changes in cholinesterase levels could not alter the sensitivity to pilocarpine which is not a substrate for the enzyme.

h) Ganglia

Many workers have sought to determine whether supersensitivity develops in the superior cervical ganglion of the cat or rat after the preganglionic fibers have been cut. Earlier studies indicated that supersensitivity to acetylcholine did occur under these conditions (see CANNON and ROSENBLUETH, 1949). Some of the more recent studies have supported this conclusion (CHIEN, 1960; BOKRI et al., 1963). On the other hand, some investigators concluded that there was no change in sensitivity of the ganglion (BROWN, 1966; VICKERSON and VARMA, 1969), a decrease in sensitivity (VOLLE and KOELLE, 1961) or perhaps a specific increase in sensitivity of muscarinic responses of the ganglion (JONES, 1963; VOLLE, 1966).

The problems in interpreting the results of experiments testing the sensitivity of the denervated superior cervical ganglion are great. For detailed consideration the reader is referred to the excellent discussions by VOLLE (1966) and GREEN (1969). Briefly, some of these problems are (1) the supersensitivity of the nictitating membrane itself, (2) the presence of tone in the supersensitive nictitating membrane, (3) decreases in cholinesterase in the denervated ganglion, (4) the existence of both nicotinic and muscarinic receptors in the ganglion, (5) a possible alteration of the relative importance of nicotinic and muscarinic receptors in the ganglion after denervation, and (6) the possibility of some preganglionic actions of cholinergic drugs. In an extensive study which attempted to deal with all these problems, GREEN (1969) came to the conclusion that the chronically denervated superior cervical ganglion of the cat is, in fact, *subsensitive* to both nicotinic and muscarinic drugs as well as to potassium. This conclusion appears to be the most supportable one at the present time.

C. Mechanisms

1. Skeletal Muscle

Denervated skeletal muscle undergoes an enormous increase in sensitivity to cholinergic substances. As discussed in Section I, there is excellent evidence in support of the conclusion that this type of supersensitivity is primarily the result of spread of cholinergic receptors outward from the endplate to make the entire fiber responsive to cholinergic agents.

However, there are other pharmacological changes in denervated muscle which cannot explained solely on the basis of cholinergic receptors. GUTMANN and SANDOW (1965) found that caffeine had the capacity to cause contracture in denervated but not control muscles. This effect of caffeine first appeared 24 hours after denervation. At that time the contracture was characterized by a long latent period. This latent period became progressively shorter as the length of time after denervation increased up to 6 days. GUTMANN and SANDOW (1965) concluded that caffeine has a calcium-releasing effect on sarcoplasmic reticulum in denervated muscle, an effect absent in normally innervated muscle. Denervated skeletal muscle also contracts in response to doses of histamine, 5-HT and bradykinin, which do not cause contraction of innervated muscle (ALONSO-DE FLORIDA et al., 1965 a, b), and is supersensitive to potassium (MAGLADERY and SÓLANDT, 1942; FLEMING, 1971).

It is thus appropriate to look for changes in the physiology of denervated skeletal muscle in addition to spread of cholinergic receptors in order to explain these altered responses to pharmacological substances. One must approach this question with caution because of the process of atrophy which also occurs as a consequence of denervation of skeletal muscle. An electron microscopic investigation by PELLEGRINO and FRANZINI (1963) has indicated that morphological signs of degeneration are seen within two weeks after the denervation of skeletal muscle. If it is assumed that supersensitivity and atrophy are separate processes, it becomes difficult to determine which sub-cellular processes are associated with supersensitivity, atrophy or both. One hopes that, by concentrating on the first few days following denervation, one is studying events related primarily to supersensitivity.

There are pronounced changes in the electrical properties of skeletal muscle cells detectable within a few days after denervation. The resting membrane potential is decreased by 10 to 15 mV, that is, the potential difference across the membrane is reduced (KLAUS et al., 1960; THESLEFF, 1963; LENMAN, 1965; ALBUQUERQUE and THESLEFF, 1968; REDFERN and THESLEFF, 1971) and spontaneous oscillations of the resting potential can be observed (LI, 1960; THESLEFF, 1963). Other alterations in electrical properties after denervation include the appearance of a large and long-lasting positive afterpotential following electrical stimulation, a decrease in potassium conductance; increases in membrane resistance, capacitance, time constant and length constant; a decrease in the rate of rise of the action potential and an increase in the duration of the action potential (HUBBARD, 1963; THESLEFF, 1963; ALBUQUERQUE and THESLEFF, 1968; REDFERN and THESLEFF, 1971). ALBUQUERQUE and THESLEFF (1968) also observed a small increase in electrical threshold.

REDFERN et al. (1970) discovered that, as the supersensitivity of denervated muscles develops, there is an increasing resistance of the action potential to tetrodotoxin. Nevertheless, the action potential is still sodium-dependent. These authors concluded that denervation had altered the spike-generating mechanism in the cells. ALBUQUERQUE and WARNICK (1972) confirmed the resistance of denervated muscle to tetrodotoxin and also observed that reducing the sodium concentration in the bathing medium partially depolarized normal muscle but hyperpolarized denervated muscle.

REDFERN and THESLEFF (1971) made a particularly interesting observation. Some of the electrophysiological changes, including the decrease in resting membrane potential, have developed fully by the second day after denervation. This is a time at which the spread of receptors is just beginning. Such a temporal dissociation of the two phenomena makes it unlikely that the electrical changes are secondary consequenecs of the spread receptors.

There are also changes in the movement and binding of ions in denervated skeletal muscles. For example, potassium permeability is decreased (KLAUS et al., 1960), and sodium permeability is increased (CREESE et al., 1968).The latter authors calculated that the changes in permeability to sodium and potassium are adequate to explain the decreased resting membrane potential. ISAACSON and SANDOW (1967) have reported that, 6 days after denervation, the influx and efflux of radiocalcium under the influence of caffeine were much greater than in control preparations. BRODY (1966) found an increase in the binding of calcium to microsomes beginning 2 to 4 days after denervation and reaching a maximum on the 15th to 17th day.

It may be concluded, therefore, that in addition to the spread of receptors, denervated skeletal muscles undergo changes in permeability and electrical properties of the membrane as well as changes in the excitation-contraction coupling mechanism. These events may well explain the supersensitivity to noncholinergic agents such as potassium ion.

2. Smooth Muscle

KIRPEKAR et al. (1962) and CERVONI and KIRPEKAR (1966) suggested that supersensitivity produced by reserpine or decentralization was due, at least in part, to altered binding and/or removal of amines from their site of action. Attractive as this proposal may have been at the time, it must now be abandoned because: (1) the supersensitivity is nonspecific, (2) reserpine does not, by itself, alter the concentration of amines at the effector site (see Section II), (3) decentralization does not affect uptake and retention in adrenergic nerve terminals (CERVONI et al., 1970).

BITO and DAWSON (1970) have proposed that postjunctional changes in sensitivity in smooth muscle are the result of an increase in the concentration of receptors. It must be pointed out that there is at present no evidence for this hypothesis. The only argument which can be advanced in favor of the theory is the fact that the number of cholinergic receptors does increase in denervated skeletal muscle. As will be pointed out later, there are important differences between skeletal muscle and smooth muscle in regard to the specificity of supersensitivity in each. Furthermore, the differences in morphology and physiology between smooth muscle and skeletal muscle make it quite unnecessary to assume that the mechanisms of supersensitivity must be identical in all respects in the two tissues. In contrast to skeletal muscle, smooth muscle cells are much smaller, have intracellular electrical contacts, are variably and diffusely innervated *en passage*, and lack both endplates and a well defined sarcoplasmic reticulum (BURNSTOCK, 1970). Furthermore, the action of potential in smooth muscle is, in part, a calcium current (TOMITA, 1970).

Skeletal muscle and smooth muscle differ greatly in regard to the specificity of postjunctional supersensitivity. In skeletal muscle the increase in sensitivity to acetylcholine is of the order of 1 000-fold whereas the increase in sensitivity to potassium is less than 2-fold (FLEMING, 1971). Receptor spread is related to this highly selective supersensitivity to cholinergic agonists. As can be seen from Section III. B and Table 1, the situation is far different in smooth muscle and cardiac muscle. The sensitivity to ions is increased 1.5- to 7-fold, depending on the tissue. The sensitivity increase to other agents, including natural transmitters and other agonists acting on receptors, is in many cases the same and never more than one order of magnitude greater in any one tissue. Thus, in smooth muscle and cardiac muscle, postjunctional supersensitivity is even quantitatively nonspecific.

Such nonspecificity argues strongly against postjunctional supersensitivity in smooth muscle being primarily a receptor phenomenon. Although the possibility of an increased number of receptors cannot be tested in smooth muscle with the methods currently available, it is possible to determine whether there have been qualitative changes in receptors. This question has been definitely answered in the negative. Experiments with the supersensitive nictitating membrane and aorta (GREEN and FLEMING, 1967; TAYLOR and GREEN, 1971) have established that the affinity of alpha-adrenergic receptors for drugs was not changed.

TAYLOR and GREEN (1971) further reported that, although the dissociation constant of the receptors for phenylephrine was not changed, the efficacy was. An increase in efficacy would be expected regardless of whether there were an increase in receptors or a change beyond the receptors such that a greater response would occur for any given number of receptors activated. This increase in efficacy adequately explains the decreased effectiveness of phenoxybenzamine in producing adrenergic blockade in the presence of postjunctional supersensitivity in the nictitating membrane (VARMA, 1966; GREEN and FLEMING, 1967; LANGER and TRENDELENBURG, 1968).

The finding of GREEN and FLEMING (1967) and TAYLOR and GREEN (1971) that postjunctional supersensitivity does not alter the pA_2 or K_B value of phentolamine is in conflict with the results of VARMA (1966) who reported a decreased effectiveness of phentolamine in denervated nictitating membranes of spinal cats. This may be due in part to the problem of tone in VARMA's experiments (LANGER and TRENDELENBURG, 1968). It is also likely that pA_2 and K_B values represent more accurate estimates of receptor affinity, especially when obtained *in vitro* as in the experiments of TAYLOR and GREEN (1971).

CERVONI et al. (1970) have suggested that postjunctional supersensitivity in the *in vivo* nictitating membrane may be related to vasodilation of the blood vessels leading to the muscle, thus allowing higher concentrations of bloodborne agents to reach the effector cells. Although such a factor may play a small role in the shifts of dose-response curves in some instances, three factors argue against its being of major importance. First, reserpine produces no supersensitivity of the nictitating membranes after one or two days of large doses which would be expected to cause marked depletion of noradrenaline stores (FLEMING and TRENDELENBURG, 1961; TRENDELENBURG and WEINER, 1962), and thus vascular dilation by removing neurogenic control (LEE, 1967). Second, decentralization supersensitivity develops slowly, whereas the vasodilatation produced by decentralization should be fully developed within a few minutes after section of the nerve. Third, postjunctional supersensitivity can be demonstrated in a variety of isolated organs in which vascular access to the cells cannot be involved.

Any mechanisms which are proposed to explain postjunctional supersensitivity must take into account the following clearly established facts which have been discussed in detail earlier in this review. (1) The supersensitivity is induced by a variety of procedures which chronically interrupt the contact between a neurotransmitter and its effector cells. (2) The phenomenon develops slowly and reaches a maximum days or weeks after the interruption is first established. (3) The supersensitivity is nonspecific. (4) It occurs in a wide variety of tissues. (5) It can be demonstrated in several tissues *in vitro*.

Consideration of the above factors led FLEMING (1963 b) to propose that postjunctional supersensitivity is the consequence of some physiological change in smooth muscle, beyond the receptors, such as an alteration in membrane permeability or in the contractile mechanism. A similar conclusion had been reached by CANNON and ROSENBLUETH (1949).

Thus far little evidence has been accumulated to indicate whether or not there is a change in the contractile apparatus. WESTFALL et al. (1972) have reported that there is no change in the amount of total protein in a supersensitive vas deferens. This does not eliminate the possibility of an undetectable increase in one particular protein. However, the conclusion that postjunctional supersensitivity is not the result of an increase in contractile protein is supported by the fact that the maximum response of tissues is not increased in postjunctional supersensitivity.

In this regard, TAYLOR and GREEN (1971) have reported an increase in the maximum response to acetylcholine in the supersensitive rabbit aortic strip. However, this is not a change in the maximum tension the tissue can develop. There was no increase in the maximum response to more powerful agonists such as phenylephrine and potassium. The change merely makes the maximum response to acetylcholine more nearly approach the maximum response to the full agonists.

The maximum response of the guinea pig vas deferens is increased by denervation (WESTFALL et al., 1972). This effect is a mere corollary to postjunctional supersensitivity because, in the same tissue, chronic pretreatment with reserpine or decentralization produces supersensitivity without an increase in maximum (WESTFALL, 1970a; WESTFALL et al., 1972). WESTFALL and coworkers therefore suggested that denervation of the vas deferens may cause, in addition to pre- and postjunctional supersensitivity, an improved electrical synchronization of the smooth muscle cells.

An increasing body of evidence is accumulating to indicate that an ionic and/or membrane mechanism is responsible for postjunctional supersensitivity in smooth muscle. The earliest evidence was of an indirect nature. ROTHBALLER and SHARPLESS (1961) reported that the chronically denervated nictitating membrane is spontaneously active and responds to stretch with a contraction. Neither phenomenon was seen in acutely denervated (control) nictitating membranes. Similarly, GREEN and FLEMING (1967) observed spontaneous activity in chronically denervated and decentralized nictitating membranes and those of cats chronically pretreated with reserpine. The response to stretch and the spontaneous activity were not prevented by section of cranial nerves, alpha adrenergic blocking agents, atropine, ganglionic blocking agents, mepyramine or procaine. They thus seem to be of myogenic origin and probably a sign of an unstable cell membrane.

Also consistent with a membrane locus for postjunctional supersensitivity are the results obtained in studies of depolarized tissues. EVANS et al. (1958) demonstrated that smooth-muscle tissue placed in a depolarizing solution would still respond to many drugs in the absence of a measurable resting membrane potential. FLEMING (1968) applied this technique to the problem of supersensitivity. He demonstrated that depolarization of control and supersensitive pieces of ileum caused the difference in their sensitivities to disappear.

In the last two years, some interesting work on the possible role of ions, especially calcium, has been published. WESTFALL (1970b) found no changes in total tissue content of sodium, potassium, calcium, magnesium or chloride in chronically decentralized or reserpine-pretreated vasa deferentia of guinea pigs. WESTFALL (1970b) also investigated the intracellular content of sodium, potassium and chloride and the rate of efflux of ^{42}K and ^{36}Cl. There were no changes in any of these parameters.

Changes also fail to appear in ^{45}Ca tissue space or total calcium content in the rabbit aorta made supersensitive by pretreatment with reserpine, as long as excessive doses of reserpine are avoided (HUDGINS and HARRIS, 1970; GARRETT and CARRIER, 1971). However, there are changes in calcium binding. HUDGINS and HARRIS (1970) noted two important changes in supersensitive aortic strips. (1) A facilitated rate of efflux of ^{45}Ca over a time period which suggested a reduced affinity of binding sites for calcium at the membrane. (2) An increase in total calcium retained by the tissue in calcium-free solution. The latter observation presumably indicates an increase in intracellular binding of calcium.

GARRETT and CARRIER (1971) confirmed the observation of an increase in total Ca^{++} retention. They also observed that supersensitivity induced by pretreatment with reserpine *reduced* the extracellular calcium requirement for responses to noradrenaline but *increased* the requirement of extracellular calcium for contractions induced by potassium chloride.

Using microelectrodes, FLEMING (1972) has begun an investigation of the electrophysiological characteristics of supersensitive cells. He has found that the vas deferens of the guinea pig, decentralized or denervated one week before the experiment, has a significantly reduced resting membrane potential.

Regardless of which procedure is used to induce supersensitivity, the potential difference across the cell membrane is about 10 mV less than control. An interesting related finding is that the denervated vas deferens is supersensitive to direct electrical stimulation (McClure et al., 1971).

Thus there seem to be changes in calcium binding, perhaps at two different sites, and a reduction in resting potential associated with postjunctional supersensitivity. Calcium is believed to play a very major role in the electrophysiology of smooth muscle cells. Calcium has a stabilizing function at the membrane (KURIYAMA, 1970) and appears to affect both sodium and potassium conductance. The resting membrane potential depends, in part, on the amount of calcium bound to the membrane. Furthermore, calcium is believed to carry at least part of the current of the action potential (TOMITA, 1970). The altered calcium binding and membrane potential in supersensitive smooth muscle may therefore be closely related phenomena. As in all muscle, calcium also plays a role in excitation-contraction coupling (BOHR, 1964).

Various drugs depend on different pools of calcium (HINKE, 1965) to produce their effects. Recently, HUDGINS and WEISS (1968) have presented evidence that potassium, noradrenaline and histamine interact with calcium by different mechanisms in aortic strips. The most common means of stimulating smooth muscle cells is via depolarization of the membrane. However, depending on the conditions and the tissue being studied, drugs may act by either electrical and/or nonelectrical events (see the review by SOMLYO and SOMLYO, 1968). In general, stimulants, such as potassium, which act primarily by depolarization, are dependent upon loosely bound and extracellular calcium. Drugs which act partly or entirely without inducing depolarization tend to be more dependent on firmly bound calcium. These considerations have led several investigators to suggest that variations among agonists in their relative dependence upon electrical events, loosely bound and firmly bound pools of calcium for producing responses may explain variations observed in the magnitude of supersensitivity demonstrated by an individual tissue to the different agonists. This concept has been discussed in detail by WESTFALL (1970a) and supported by HUDGINS and HARRIS (1970), GOODMAN and WEISS (1971), GARRETT and CARRIER (1971) and CARRIER and JUREVICS (1972).

3. Cardiac Muscle

Morphologically and physiologically cardiac muscle is intermediate between skeletal muscle and smooth muscle. However, postjunctional supersensitivity in cardiac muscle has the same characteristics as in smooth muscle. That is, in contrast to skeletal muscle, the sensitivity is increased rather equally to neurotransmitters and ions (WESTFALL and FLEMING, 1968a, b). It is reasonable at this stage to anticipate that the mechanism of the supersensitivity is similar in smooth muscle and cardiac muscle.

There is indirect evidence in support of a membrane locus for supersensitivity in hearts. As discussed in Section III. B, postjunctional supersensitivity to chronotropic and arrhythmic actions of drugs has been observed. Rate and rhythm are clearly membrane-regulated functions. The small amount of evidence thus far available indicates that there may not be a similar supersensitivity to inotropic effects. There is supersensitivity to the chronotropic effects of calcium (WESTFALL and FLEMING, 1968a, b). Just as in smooth muscle, calcium plays an important role in both membrane stability and the action potential (see, for example, CARMELIET and VEREECKE, 1969). There are, at present, no studies of calcium binding and movement or of electrophysiologic characteristics of supersensitive hearts.

MCNEILL (1969, 1970) has reported that reserpine, 2.5 to 5.0 mg/kg, 24 hours before the experiment, enhances the activation of phosphorylase produced by noradrenaline and isoproterenol in the rat heart. The relationship of this important observation to postjunctional supersensitivity must be considered with caution, however, since a smaller dose of reserpine (0.5 mg/kg/day) for up to 12 days did not alter the ability of noradrenaline to activate phosphorylase. One would have expected adequate depletion and postjunctional supersensitivity with MCNEILL's lower dose. With daily doses of 0.05 to 0.1 mg/kg for 1 to 10 days, KOVACIC and ROBINSON (1966) and SJÖSTRAND and SWEDIN (1968) showed substantial depletion of noradrenaline stores in the rat heart.

In 1972, MCNEILL and SCHULZE demonstrated that reserpine caused a shift of the dose-response curves for both noradrenaline and histamine in activating phosphorylase in guinea pig hearts. The dose of reserpine was 3 mg/kg, 48 hours and 2 mg/kg, 24 hours before the experiment. Again, these doses are large. It is important that small doses of reserpine be tested in the guinea pig heart. For discussion of the importance of dose to the effects of reserpine, see Section II. The possible role of phosphorylase in supersensitivity should be investigated further because of the relationship of calcium to phosphorylase activation and the evidence (vide infra) of changes in adenyl cyclase in the denervated pineal gland.

4. Nonmuscular Tissues

Direct evidence of possible mechanisms of supersensitivity in tissues other than muscle is scarce. WEISS and COSTA (1967) demonstrated that chronic denervation markedly increased the activation by noradrenaline of adenylcyclase in the rat pineal gland. Subsequently, WEISS (1969) reported that chronic denervation of the pineal potentiated the activation of adenylcyclase produced by both noradrenaline and sodium fluoride. The evidence indicated an increase in the amount of the enzyme present.

BURFORD and GILL (1961) investigated the role of calcium in the function of normal and supersensitive submaxillary glands of the cat. They found no change which would explain the supersensitivity. Studies of salivary glands with microelectrodes have indicated that the resting potential of acinar cells is normally quite low and that the response to stimulation of either the parasympathetic or sympathetic nerves leads to hyperpolarization in some cells and depolarization in others. There is as yet no evidence for or against a role of membrane electrical properties in supersensitive salivary glands. In view of the unpredictable nature of the electrophysiological responses, such evidence may be very difficult to obtain.

D. Supersensitivity in Isolated Tissues as a Function of Mechanisms

In 1968, TSAI et al. reported the results of an extensive study of supersensitivity in the isolated nictitating membrane of the cat. They showed conclusively that chronic procedures which produced postjunctional supersensitivity of the *in vivo* nictitating membrane did not do so when the sensitivity was tested *in vitro*. The procedures included denervation, decentralization, the chronic administration of reserpine and chronic administration of a ganglionic blocking agent. TSAI et al. did note the development of spontaneous activity after these procedures, suggesting some kind of postjunctional change. Prejunctional supersensitivity was readily demonstrable *in vitro*. TRENDELENBURG et al. (1970) also found no sign of postjunctional supersensitivity in the isolated nictitating membrane.

Postjunctional supersensitivity was also undetected in the isolated spleen of the cat (GREEN and FLEMING, 1968), in the isolated atrium of the guinea pig (WESTFALL and FLEMING, 1968a), rabbit (SHIBATA et al., 1972) or rat (COREY and McPHILLIPS, 1972); in the perfused portal vein of the rat (JOHANSSON et al., 1970); in the perfused mesenteric artery of the rat (HAEUSLER and HAEFELY, 1970); and in the isolated longitudinal muscle of the guinea pig ileum (D. P. WESTFALL, unpublished results). Postjunctional supersensitivity has been detected *in vivo* in some of these tissues; viz. cat spleen (BURN and RAND, 1959), guinea pig heart (WESTFALL and FLEMING, 1968a) and rat heart (COREY and McPHILLIPS, 1972).

TSAI et al. (1968) concluded that postjunctional supersensitivity was a phenomenon demonstrable primarily by experiments in which non-steady-state responses are studied. According to TSAI and his coworkers, postjunctional supersensitivity does not appear *in vitro* because steady-state responses are obtained *in vitro*. By steady-state responses, TSAI and his coworkers imply responses which rise to a plateau level and remain at that level. They do acknowledge that other factors, such as trauma in isolation and changes in ion balance during isolation may be masking postjunctional supersensitivity as suggested by WESTFALL and FLEMING (1968a).

TSAI (1970) has presented good evidence that the magnitude of postjunctional supersensitivity in the *in vivo* nictitating membrane is reduced (but not abolished) under steady-state conditions of drug administration (i.e. when the drug is infused rather than injected). However, if one takes the broad view of postjunctional supersensitivity, any conclusion that the phenomenon is simply an *in vivo* or a non-steady-state phenomenon is untenable. The list of isolated tissues which have failed to show postjunctional supersensitivity is equaled or exceeded by the list of isolated tissues in which it has been demonstrated. These include: rabbit aortic strips, guinea pig ileum, guinea pig and rat vas deferens, perfused mesenteric artery of dogs, perfused ear artery of rabbits, perfused guinea pig heart, iris sphincter of the cat and pig, and the dog heart-lung preparation (semi-isolated). See Section III. B for details and references.

Furthermore, postjunctional supersensitivity can be demonstrated *in vitro* with cumulative dose-response curves (for example, in aortic strips by HUDGINS and FLEMING, 1966; TAYLOR and GREEN, 1971; MALING et al., 1971 and SHIBATA et al., 1972). Cumulative dose-response curves *in vitro* result in steady-state responses.

WESTFALL and FLEMING (1968a) presented another possible explanation for the fact that some isolated tissues exhibit postjunctional supersensitivity and some do not. This hypothesis is based on the recognized effect of isolation procedures on ion balance in tissues (BOHR, 1964). Ions tend to move down along their concentration gradients immediately after isolation and then gradually return toward, but not always reach, normal concentrations. Furthermore, this breakdown in ion balance appears to be a function of the handling of the tissue during isolation (DAWKINS and BOHR, 1960). For example, BAUER et al. (1965) have found that manipulation or damage to the taenia coli significantly increases the calcium content. WESTFALL and FLEMING (1968a) therefore suggested that, since postjunctional supersensitivity appears to involve such factors as membrane potential, ion fluxes and ion binding in the tissue, upsetting the ion balance could mask the phenomenon. Consistent with this hypothesis is the fact that the isolation of the nictitating membrane requires an unusual amount of manipulation of the tissue.

CERVONI et al. (1970) argue against the trauma hypothesis. According to them, if the isolated nictitating membrane has been affected by ion changes associated with trauma, it is inconsistent that (1) uptake of noradrenaline into neurons and (2) maximum tension development are "unimpaired". These arguments are easily countered. (1) There is no reason why the state of neurons should reflect the state of smooth muscle. Furthermore, isolation and the associated ion changes have been shown to be associated with changes in neuronal retention of ^{3}H-noradrenaline (Westfall and Peach, 1970). (2) Complete depolarization of smooth muscle by replacing the Na$^+$ and Cl$^-$ in the medium by K$^+$ and SO$_4^-$ does not necessarily reduce its maximal contractile ability (Evans et al., 1958). Thus, maximum contractile force seems to be relatively independent of changes in concentration of at least some ions.

Several lines of evidence have been found in our laboratory to support the trauma hypothesis of Westfall and Fleming (1968a). In the above publication it was noted that chronic administration of reserpine to guinea pigs produced a rather large increase in sensitivity to the chronotropic effects of noradrenaline *in vivo*. The shift of the dose-response curve was less, but still significant, if measured in perfused hearts but it could not be demonstrated in the isolated atrium. Westfall and Fleming proposed that the perfused heart and isolated atrium represented lesser and greater degrees respectively of trauma in the isolation procedure.

Subsequently, Westfall and Fleming (1968c) studied supersensitivity in perfused rabbit hearts. They reasoned that, if the imbalance in ions were pronounced soon after isolation and gradually returned toward normal (Bohr, 1964) and if this imbalance were masking supersensitivity, then the duration of isolation might show a relationship to the magnitude of the difference in sensitivity between control and treated tissues. The results supported this assumption. Control hearts and hearts from rabbits pretreated with reserpine presented identical sensitivities to noradrenaline measured after one hour of perfusion. However, after two hours of perfusion, the treated hearts were significantly more sensitive than controls.

A final piece of evidence comes from experiments with the guinea pig ileum. Fleming (1968) established that chronic ganglionic blockade in guinea pigs produced postjunctional supersensitivity of the ileum, tested *in vitro*. However, D. P. Westfall (unpublished results) observed that identical experiments with isolated strips of longitudinal smooth muscle of the ileum yielded no sign of supersensitivity. In contrast to the results with "sleeves"

Table 2. Longitudinal muscle of guinea-pig ileum

Agonist	Group	Geometric mean ED 50 (95% CI) ($\times 10^{-7}$ M)	N
ACh	Saline	3.34 (1.51–7.56)	5
	Chlorisondamine	3.40 (0.88–13.2)	5
Histamine	Saline	3.95 (1.18–13.3)	5
	Chlorisondamine	3.55 (1.69–7.50)	5

Chlorisondamine administered 1.0 mg/kg, 3 × daily for 5–7 days.
Saline administered 1.0 ml/kg, 3 × daily for 5–7 days.

of ileum (FLEMING, 1968; see Table 1, Section III. B), WESTFALL obtained the results shown in Table 2. It thus appears that the additional manipulation necessary to isolate the longitudinal strips caused the supersensitivity to disappear and/or be masked.

The ultimate proof or disproof of the trauma theory will depend upon correlations between ion determinations and supersensitivity in various isolated tissues. Experiments of that type are now under way in our laboratory.

IV. Postjunctional Subsensitivity

A. Tolerance to Cholinesterase Inhibitors

If decreasing the amount of transmitter to which an organ is exposed brings about supersensitivity of the effector to chemical stimuli then an increase in the amount of transmitter should be expected to cause a diminished responsiveness of the effector, or subsensitivity. The phenomenon of subsensitivity does occur but has not been studied as extensively as supersensitivity.

The first indication of subsensitivity came from a study by RIDER et al. (1952), who were evaluating the chronic toxicity of octamethyl pyrophosphoramide (OMPA), an organophosphorus cholinesterase inhibitor. When given to rats in daily doses of 1.5–2.0 mg/kg OMPA caused 100 % mortality after 5 or 6 doses. However, if the rats were started at a dose of 0.5 mg/kg and the dose gradually increased it was possible to build tolerance to OMPA to the extent that the rats could eventually withstand a dose of 1.5 mg/kg for more than 20 days. In a later study BARNES and DENZ (1954) demonstrated a very high degree of tolerance to O-O-diethyl O-2-ethylmercaptoethyl thionophosphate (Systox), which is also a cholinesterase inhibitor. Systox was fed to rats in the diet and at the end of one year the rats were consuming the equivalent of a single lethal dose of the compound each day. Other than a slight transient weight loss the rats showed no ill effects. Brain cholinesterase, however, was inhibited to the extent of 95 %. There are also reports of tolerance to other organophosphates such as disulfoton (BOMBINSKI and DuBois, 1958), parathion (OLIVER and FUNNELL, 1961) and diisopropyl fluorophosphate (DFP) (JOHNS and McQUILLEN, 1966).

The action of the organophosphates is the result of cholinesterase inhibition and the subsequent accumulation of acetylcholine at junctional sites. Tolerance to cholinesterase inhibitors, therefore, suggests an accommodation or adaptation to the effects of acetylcholine. This idea is supported by observations made by BRODEUR and DuBois (1963). They carried out experiments to determine the mechanism of tolerance to disulfoton. Rats were treated with a fixed dose of disulfoton for 30 days. After approximately 3 days of treatment the rats developed the signs which are characteristic of intoxication with cholinesterase inhibitors viz.; excessive salivation and lacrimation, frequent urination, diarrhea, tremors and muscle fasciculations. At about the tenth day of treatment the signs of poisoning began to subside and by the thirtieth day the intensity of the signs was considerable diminished. At this point the toxicity of carbachol, a nonhydrolyzable cholinester, was measured in treated and control rats. The rats which had developed tolerance to disulfoton were found to be only half as sensitive as control rats to the lethal action of carbachol. The LD 50 of carbachol increased from 2.0 mg/kg in the control group to 3.9 mg/kg in the tolerant group. Thus, tolerance to disulfoton was accompanied by subsensitivity to carbachol. The results suggested that perhaps there was a decrease in the sensitivity of the effector cells to carbachol and possibly to other agonists which act at the same receptor.

B. Subsensitivity in Various Organs and Tissues

At the same time as BRODEUR and DuBois (1964) reported subsensitivity to the lethal effect of carbachol, EMMELIN (1964) reported subsensitivity of the salivary gland of the cat. Subsensitivity has subsequently been detected in a number of different preparations both *in situ* and *in vitro*. Table 3 lists the organs and tissues in which subsensitivity has been studied.

1. Exocrine Glands

The salivary gland is an organ which has been studied extensively with respect to changes in sensitivity to chemical stimuli. The work on the salivary gland was reviewed by EMMELIN in 1961. EMMELIN and his coworkers have shown that when the gland is disconnected from the central nervous system, either by surgical means or by drugs, the organ is deprived of its transmitter and supersensitivity develops. If, on the other hand, the salivary gland is exposed to excessive amounts of the transmitter the gland becomes subsensitive (EMMELIN, 1964). Cats were treated with physostigmine twice daily for two days in order to produce excess amounts of acetylcholine. Twenty-four hours after the last dose of physostigmine the sensitivity of the submaxillary gland was tested by administering adrenaline to evoke salivary secretion. EMMELIN found that in both innervated and decentralized glands there was a reduced sensitivity to adrenaline following subacute administration of physostigmine.

The sweat glands of the cat have also been shown to exhibit subsensitivity. REAS and TRENDELENBURG (1967) were interested in comparing supersensitivity caused by denervating an adrenergically innervated tissue (nictitating membrane) with supersensitivity caused by denervating a cholinergically innervated tissue (sweat glands). In the course of these experiments they discovered that during the first two days after denervation the sweat glands

Table 3. Subsensitivity in various tissues and organs

Species	Tissue	In vitro	Treatment Type	Days	Agonist	Shift	Reference
Rat	[a]	—	disulfoton	30	carbachol	2	Brodeur and DuBois (1964)
	[a]	—	disulfoton	30	carbachol	1.3	McPhillips and Dar (1967)
	heart[b]	—	disulfoton	10	carbachol	2.5	Perrine and McPhillips (1970)
	atrium[b]	yes	disulfoton	8	carbachol	2.1	Perrine and McPhillips (1970)
	atrium	yes	disulfoton	8	bethanechol	3.3	Perrine and McPhillips (1970)
	ileum	yes	disulfoton	10	carbachol	3.4	McPhillips (1969)
		yes	disulfoton	10	oxotremorine	3.0	McPhillips (1969)
		yes	disulfoton	10	furthrethonium	2.8	McPhillips (1969)
		yes	disulfoton	10	arecoline	3.3	McPhillips (unpublished)
Cat	submaxillary gland	—	physostigmine		adrenaline	[c]	Emmelin (1964)
	sweat gland	—	Den	2	acetylcholine	10	Reas and Trendelenburg (1967)
		—	Den	2	pilocarpine	3	Reas and Trendelenburg (1967)
	nicticating membrane	—	Den	1.5	methoxamine	1.5	Trendelenburg et al. (1970)
	iris	—	continuous light, 300 Lux	7	pilocarpine	4.1	Bro et al. (1971)

[a] This represents experiments in which the LD 50 of carbachol was measured in treated and control rats.
[b] The responses measured in these experiments were negative chronotropic responses.
[s] Responses were not reported in a manner in which shift of the dose response curve could be calculated.

were subsensitive to both acetylcholine and pilocarpine. It is interesting that subsensitivity developed as a consequence of denervation. In this instance, however, as in the previous one, subsensitivity is thought to be associated with the gland being subjected to excessive amounts of acetylcholine. REAS and TRENDELENBURG pointed out that subsensitivity to acetylcholine and pilocarpine probably coincided with release of transmitter from degenerating nerve terminals, a situation which will increase the amount of transmitter to which the gland is exposed.

Evidence for the release of acetylcholine from nerve terminals following denervation was presented by EMMELIN and STRÖMBLAD (1958) and EMMELIN (1962). In the parotid gland, for example, one day after denervation there were periods of spontaneous secretory activity which could be abolished by atropine and augmented by physostigmine. Similar observations have been made on the submaxillary and sublingual glands of the cat (EMMELIN, 1962). The conclusion, therefore, was that the episodes of secretory activity were associated with the release of acetylcholine from the terminals of the degenerating nerve fiber.

Subsensitivity following denervation, then, is associated with excessive amounts of acetylcholine. Apparently a similar situation develops in adrenergically innervated tissue immediately following denervation. As pointed out by REAS and TRENDELENBURG (1967) there is difficulty in detecting subsensitivity because of the simultaneous development of prejunctional supersensitivity which is caused by loss of the uptake mechanism. Nevertheless, TRENDELENBURG et al. (1970) were able to detect a slight subsensitivity to methoxamine in the acutely denervated nictitating membrane of the cat. Methoxamine is an adrenergic agent which is not taken up by the nerve terminal and therefore prejunctional supersensitivity does not develop to this compound.

2. Heart

In 1967 McPHILLIPS and DAR reported subsensitivity to the negative chronotropic action of carbachol in rats which had been given disulfoton for 30 days. The administration of disulfoton for this length of time did not appear to have any measurable adverse effect on the cardiovascular system. There was no difference between treated and control groups with respect to resting heart rate or blood pressure at the time the sensitivity to carbachol was tested. Nevertheless there was subsensitivity to the negative chronotropic action of carbachol; there was a 3-fold shift of the dose-response curve to the right. Resistance to cardiac arrest was another interesting finding of that study. In 10 of 12 controls rats 30 μg/kg of carbachol produced cardiac arrest. However, cardiac arrest could not be produced with 30 μg/kg of carbachol in any rats treated with disulfoton for 10 days or more. Some rats apparently had developed subsensitivity to the point where doses of carbachol as high as 300 µg/kg produced only a very slight depressor response and mild brady-cardia which was immediately followed by tachycardia and a pressor response. In some cases the blood pressure rose 95 mm Hg above the preinjection level. It was concluded that these rats had developed a very high degree of resistance to the negative chronotropic and depressor actions of carbachol but not to the action of carbachol on sympathetic ganglia or the adrenal medulla. This suggests the possibility that perhaps subsensitivity develops more readily to the muscarinic actions of carbachol than to the nicotinic effects. This question is yet to be explored, however.

Subsequent studies in which the isolated spontaneously beating atrium was used confirmed the observation that subsensitivity develops to the nega-tive chronotropic action of carbachol (PERRINE and McPHILLIPS, 1970). Atria were removed from rats which were treated for 8 days with disulfoton and compared with atria taken from control rats. The mean resting rate of atria from the disulfoton treated rats was not different from the mean atrial rate of the control group. There was however a 3-fold difference in sensitivity to the negative chronotropic action of carbachol (Table 3). There was ap-proximately the same degree of subsensitivity to bethanechol, which is also a nonhydrolyzable choline ester.

The sensitivity of the atrium to acetylcholine was markedly increased. There was approximately a 17-fold shift of the dose-response curve to the left. This was not an unexpected observation because the animals were being treated with a compound which inhibits hydrolysis of acetylcholine and in the atrium there was an 80 % reduction in the rate of acetylcholine hydrolysis (COREY, 1970). The sensitivity of the atrium to methacholine was increased approximately 4-fold.

The noncholinergic agonists veratramine and adrenaline were also tested by PERRINE and McPHILLIPS (1970) but there was no change in sensitivity to these compounds following subacute administration of disulfoton. The implication of those results are discussed below under "nonspecificity".

3. Ileum

The ileum develops changes in sensitivity almost identical to those seen in the atrium (McPHILLIPS and DAR, 1967; McPHILLIPS, 1969). As in previous experiments disulfoton was used to produce subsensitivity. The ileum was shown to be subsensitive not only to carbachol but also to oxotremorine, furtrethonium and arecoline. The change in sensitivity to each of these com-pounds was essentially the same, approximately 3-fold.

The sensitivity of the ileum to acetylcholine, butyrylcholine and metha-choline was also determined (McPHILLIPS, 1969). As anticipated, the actions of acetylcholine and butyrylcholine were greatly potentiated in segments of ileum taken from disulfoton-treated rats. It was surprising to find, however, that the sensitivity of the ileum to methacholine was unchanged by disulfoton administration.

Changes in the sensitivity of a tissue to hydrolyzable choline esters are difficult to interpret when the animals have previously been treated with a cholinesterase inhibitor. The fact that the actions of acetylcholine and butyrylcholine are potentiated does not mean that a tissue is not subsensitive. It could be argued that subsensitivity to hydrolyzable esters does occur but is obscured by the potentiation which occurs as a result of inhibition of hydrolysis. Since subsensitivity develops to nonhydrolyzable cholinergic agonists such as carbachol, bethanechol, etc., it seems reasonable to assume that subsensitivity develops to the hydrolyzable agonists as well. Furthermore, subsensitivity to acetylcholine is the logical explanation for the tolerance which develops to the cholinomimetic and lethal effects of cholinesterase inhibitors.

Recently, Foley and McPhillips (unpublished results) made an attempt to determine whether subsensitivity to a hydrolyzable ester could be detected. The ester chosen was methacholine. In order to complete these experiments it was necessary to compare sensitivity to methacholine in segments of ileum which were subsensitive with sensitivity to methacholine in segments which were not. It was also necessary to have cholinesterase levels depressed to the same extent in both groups. This was accomplished by administering a single dose of 1.5 mg/kg of disulfoton. When measured 18–20 h later, methacholine hydrolysis was inhibited to the extent of 54%. Simultaneous measurement of methacholine hydrolysis in segments of ileum which were subsensitive (i.e. taken from rats treated for 8 days with disulfoton) revealed 52% inhibition of hydrolysis. Methacholine hydrolysis was therefore inhibited to the same extent in both groups. The sensitivity of the ileum to methacholine is shown by the data in Table 4. A single dose of disulfoton, as expected, caused an increase in the sensitivity of the ileum. The ED 50 decreased from 0.51 to 0.34 µM. When the rats were treated for 8 days with disulfoton there was

Table 4. Sensitivity of rat ileum no methacholine after acute and subacute treatment with disulfoton

Treatment	Group	N[a]	ED 50 µM	95% Confidence limits
Disulfoton 1.5 mg/kg 1 day	Control	15	0.51	0.44–0.61
	Treated	13	0.34[c]	0.27–0.44
Disulfoton 1.0 mg/kg 8 days	Control	5	0.69	0.42–1.12
	Treated	5	0.77	0.49–1.19

[a] N indicates the number of rats per group.
[b] Geometric mean.
[c] Significantly different from controls at the 0.01 level of probability.

no difference between treated and control groups. The potentiation observed after the single injection was no longer present, indicating that subsensitivity to methacholine had developed.

Attempts to detect subsensitivity of the ileum to noncholinergic agonists were not successful. The ileum does not develop subsensitivity to potassium ion (McPHILLIPS, 1969). This observation was confirmed in later experiments (FOLEY and McPHILLIPS, unpublished results) in which adjacent segments of ileum were taken from the same rat and tested simultaneously; one with carbachol and the other with potassium. Subsensitivity did develop to carbachol, but segments of ileum from the same rat showed no change in sensitivity to potassium. The implications of these findings are discussed below under "nonspecificity".

4. Iris

Subsensitivity to cholinergic agents has been reported by BITO et al. (1967). When DFP and echothiophate are administered topically they lose their effectiveness as miotic agents after several days of application. The degree of subsensitivity produced by local application of the cholinesterase inhibitors is quite marked. After 6 days of treatment with echothiophate the ability of a 1.5 % solution of carbachol to constrict the pupil of the dog was essentially abolished. The same was true of pilocarpine. In fact, in the eye treated subacutely with echothiophate, pilocarpine was converted to a mydriatic.

Subsensitivity is not the result of damage to the pupil, because the pupils showed normal or hyperactive responses to light. Response to quantitative changes in light were not investigated, however. BITO and coworkers also pointed out that subsensitivity was not the result of a decreased permeability of the outer coat of the eye to the drugs. This possibility was eliminated by the injection of carbachol directly into the vitreous humor. Even when drugs were injected by this route there was still marked subsensitivity.

Physostigmine was also used by BITO and coworkers but they found it to be much less effective than echothiophate in producing subsensitivity to carbachol. They point out, however, that the inability of physostigmine to render the eye completely insensitive to carbachol is probably related to the short duration of action of this cholinesterase inhibitor. The suggestion is that in order for subsensitivity to develop, cholinesterase must be depressed and remain depressed for relatively long periods of time.

In a more recent study, BITO et al. (1971) produced changes in the sensitivity of the iris by altering the intensity of the physiological stimulus, light. Keeping the animals in complete darkness, a situation which reduces the cholinergic input to the iris, caused supersensitivity to the miotic effect of pilocarpine, Exposing the animals to continuous light, a situation which increases the cholinergic input to the iris, leads to subsensitivity to the miotic action of pilocarpine.

The examples of subsensitivity discussed above suggest that the sensitivity of the effector is inversely related to the concentration of the neurotransmitter to which the effector is exposed. This hypothesis was proposed by Emmelin (1964) and restated later by Bito et al. (1971), who also propose that supersensitivity and subsensitivity are opposite expressions of one basic phenomenon. This is an attractive hypothesis because there are several characteristics which are common to supersensitivity and subsensitivity.

C. Characteristics and Possible Mechanisms of Subsensitivity
1. Time Course

Subsensitivity, like supersensitivity, requires time to develop. Apparently the effector must be exposed continuously to increased levels of the transmitter for a relatively long period of time since subsensitivity does not develop following single doses of a cholinesterase inhibitor even when the dose is sufficiently high to cause acute signs of poisoning (McPhillips, 1969; Perrine and McPhillips, 1970). Longer periods of exposure are required. Emmelin (1964) detected subsensitivity in the cat only after 2 days of exposure to physostigmine and McPhillips and Dar (1967) did not detect subsensitivity to carbachol in the ileum until after 3 days of treatment. Similar results have been reported for the atrium (Perrine and McPhillips, 1970) and the iris (Bito et al., 1967).

Subsensitivity is reversible. Normal sensitivity returns when administration of the cholinesterase inhibitor is stopped. Foley and McPhillips (unpublished results) found that in the ileum normal sensitivity to carbachol returned 8 days after administration of disulfoton was discontinued. Cholinesterase levels in the ileum, however, were back to normal within 5 days after treatment was stopped. Presumably subsensitivity persists beyond the time when acetylcholine levels return to normal.

The cellular or subcellular changes which account for subsensitivity, therefore, are slow to develop and slow to disappear.

2. Postjunctional Site of Subsensitivity

Supersensitivity which develops when an effector is disconnected from the central nervous system is postjunctional in origin. It appears that subsensitivity is also postjunctional in origin. This conclusion is reached primarily by eliminating the nerve terminal as a possible site of subsensitivity. In 1961 Volle and Koelle proposed that carbachol acted indirectly, by causing the release of acetylcholine from nerve terminals. One might argue therefore that all the agents to which subsensitivity develops also exert their effect indirectly by causing acetylcholine release and that subsensitivity occurs because the release of acetylcholine is impaired. It seems unlikely that this is the mechanism of subsensitivity. It is well established that when cholinesterase is inhibited the action of acetylcholine is potentiated. A single injection of disulfoton, 1.8 mg/kg, to a rat results in a 5-fold increase in sensitivity to acetylcholine in segments of rat ileum (McPHILLIPS, 1969). If agents such as carbachol, oxotremorine, etc. caused release of acetylcholine from nerve terminals one should expect potentiation of the effect of these compounds as well. The same dose of disulfoton (1.8 mg/kg) however, did not change the sensitivity of the ileum to carbachol or oxotremorine.

The results of EMMELIN's experiments can also be used to argue that the nerve terminal is not involved in subsensitivity. It seems highly improbable that subsensitivity of the salivary gland to adrenaline (EMMELIN, 1964) can be explained by an impaired ability of the cholinergic nerve terminal to release acetylcholine since there is no evidence that the effects of adrenaline are exerted indirectly via cholinergic nerve terminals.

3. Nonspecificity

The principal characteristic of postjunctional supersensitivity is nonspecificity. It has not yet been clearly established whether subsensitivity is specific or not, because there are conflicting data concerning the nature of subsensitivity. EMMELIN (1964) treated cats with physostigmine and detected subsensitivity of the salivary gland to adrenaline. Cholinergic agonists were not tested. However, the fact that there was subsensitivity to a noncholinergic agonist suggests that the salivary gland was probably subsensitive to any agonist which would initiate a response. Attempts to demonstrate the nonspecific nature of subsensitivity in other tissues have not been successful. The rat ileum is subsensitive to several cholinergic agonists such as carbachol, oxotremorine and arecoline but not to potassium ion (McPHILLIPS, 1969; FOLEY and McPHILLIPS, unpublished results). These results are somewhat surprising because the conditions used to produce subsensitivity (increased acetylcholine levels produced by inhibition of cholinesterase) were exactly the opposite of those used by FLEMING (1968) to cause supersensitivity of the guinea pig ileum (decreased acetylcholine caused by chronic ganglion blockade), and FLEMING found supersensitivity to potassium as well as other noncholinergic agents.

The rat atrium also failed to show nonspecific subsensitivity (PERRINE and McPHILLIPS, 1970). The atrium developed subsensitivity to carbachol and bethanechol but did not develop subsensitivity to adrenaline or veratramine. Veratramine, a veratrum alkaloid, has negative chronotropic effects which do not appear to be mediated through the cholinergic receptor (LANGER and TRENDELENBURG, 1964).

Nonspecificity cannot be ruled out on the basis of the data presented by McPhillips (1967) and Perrine and McPhillips (1970). There are other points which must be considered. The rat ileum, for example, is a peculiar tissue. It responded only to cholinergic agonists and potassium ion. Histamine, 5-hydroxytryptamine and prostaglandin E_1 failed to contract the rat ileum (Foley and McPhillips, unpublished results). Even though there was no change in sensitivity to potassium it is difficult to draw a definite conclusion about specificity on the basis of results with one agonist on a tissue which is unresponsive to a wide variety of agonists.

One must also consider that the degree of subsensitivity which develops in the ileum is slight. The maximum change in sensitivity has been approximately 3-fold. It should be pointed out that changes in sensitivity are not equal for all agonists. For example, Hudgins and Fleming (1966) (Table 1) found a 17-fold increase in sensitivity of the aorta to noradrenaline but only a 1.7-fold increase in sensitivity of the same tissue to potassium ion. It is possible that subsensitivity to potassium ion does develop but is not detectable until the degree of subsensitivity to agents such as carbachol is increased to a very great degree.

A similar explanation could be offered for the failure of the atrium to show subsensitivity to veratramine. Furthermore, veratramine has some shortcomings as an agonist. It was not possible to obtain dose-response curves with veratramine because the effects on the atrium persisted for a considerable length of time after the drug was flushed from the tissue chamber. Also, the response to veratramine develops slowly; up to 20 minutes were required for the full effect to occur. Perrine and McPhillips (1970) therefore made comparisons on the basis of a response to a single concentration. The concentration chosen was one which decreased atrial rate by approximately 50%. Despite the fact that there was no subsensitivity to veratramine it is again difficult to draw definite conclusions about specificity on the basis of one agonist.

In contrast to the ileum and the atrium, the uterus of the rat is responsive to a wider variety of agonists. This organ seemed ideal for studying the problem of specificity because the rat uterus responds to angiotensin, 5-hydroxytryptamine, prostaglandin, oxytocin and potassium as well as cholinergic agents. Foley and McPhillips (1971) found, however, that subacute administration of disulfoton to rats does not produce subsensitivity to carbachol or anything else. Uterine horns were removed from rats which had been treated for 8 days with disulfoton. Segments of ileum were taken from the same rats. The uterus did not exhibit subsensitivity. In segments of ileum taken from the same rats and studied at the same time the ED 50 of carbachol increased from 0.21 µM to 0.54 µM.

The rat uterus contains a relatively sparse cholinergic innervation (ADHAM and SCHENK, 1969) and this may be the reason why subsensitivity does not develop in this tissue. If the cholinergic innervation is relatively sparse then the acetylcholine content of the tissue is probably very low. Inhibition of cholinesterase, regardless of the degree, may be insufficient to elevate acetylcholine levels in the biophase to a point which is adequate to induce subsensitivity.

Perhaps more important than the density of innervation or stores of acetylcholine is the flow of impulses to a tissue. There is little doubt that the ileum receives a continuous flow of impulses from the parasympathetic nerves. It is unlikely, however, that the uterus receives a continuous flow of impulses from cholinergic nerves. If the flow of impulses is very small then not very much acetylcholine will accumulate in the biophase even though cholinesterase is inhibited. Subsensitivity would therefore not be expected to develop.

The problem of specificity, therefore, is yet to be solved. It is probable, however, that subsensitivity is the result of a postjunctional change which produces subsensitivity to all agonists. EMMELIN's finding of subsensitivity of the submaxillary gland to adrenaline is the strongest point of evidence to support that conclusion.

4. Receptors and Subsensitivity

In trying to explain tolerance to disulfoton and subsensitivity to carbachol BRODEUR and DuBOIS (1964) proposed that the cholinergic receptor was involved. They suggested that exposure of the tissue to increased levels of acetylcholine causes the receptors to become refractory. Another possibility is that cholinesterase inhibitors may combine irreversibly with the cholinergic receptor as well as the chemically related cholinesterase enzyme. In fact, there are investigators who have proposed that the cholinergic receptor and the cholinesterase enzyme are one and the same (ZUPANČIČ, 1967). However, single doses of disulfoton (1.8 mg/kg) which caused marked inhibition of cholinesterase did not cause subsensitivity to carbachol (McPHILLIPS, 1969; PERRINE and McPHILLIPS, 1970). Moreover, subsensitivity was not detected in the uterus of rats which were treated for 8 days with disulfoton, a treatment which does cause subsensitivity of the ileum (FOLEY and McPHILLIPS, 1971). The uterus does possess cholinergic receptors, because it contracts when exposed to carbachol and acetylcholine. If disulfoton combined irreversibly with receptors in the ileum then receptors of the uterus should also have been inactivated by disulfoton.

A reduced affinity of the cholinergic receptor for drugs could also explain subsensitivity. If subsensitivity were the result of a change in the affinity of the cholinergic receptor for agonists such as carbachol and bethanechol, one might anticipate a change in affinity for an antagonist such as atropine. PERRINE and MCPHILLIPS (1970) estimated the affinity of the rat atrium for atropine by measuring pA_2 values for this antagonist in normal atria and atria which were subsensitive. Despite a 3-fold decrease in sensitivity to carbachol there was no change in the pA_2 values for atropine. The results do not support the idea that subsensitivity is related to a decreased affinity of the cholinergic receptor. It is also difficult to explain decreased sensitivity of the salivary gland to adrenaline (EMMELIN, 1964) on the basis of a change in the affinity of the cholinergic receptor for drugs.

BITO and coworkers (1970, 1971) propose that changes in the sensitivity of smooth muscle as well as other innervated target organs result from changes in the concentration of cholinergic receptors on the effector cell membrane. There is no evidence to support such a hypothesis, however, (see Section III. C). If subsensitivity and supersensitivity represent opposite expressions of a single basic mechanism, as BITO and his coworkers propose, they should share common characteristics. The preponderance of evidence presented in other sections of this review clearly excludes the receptor as the primary site of supersensitivity. The most reasonable hypothesis seems to be that supersensitivity and subsensitivity are opposite expressions of one fundamental phenomenon, that the receptor is not involved and that the changes responsible for subsensitivity are postjunctional in origin.

It is proposed that the sensitivity of effector cells is subject to a slow mechanism of adaptation: Whenever the effector cells are deprived of the transmitters, postjunctional supersensitivity develops slowly; whenever the biophase is flooded with the transmitter (or a related agent), the sensitivity of the effector cells decreases slowly. Furthermore, the set point of normal sensitivity may well differ from organ to organ, since organs which receive few autonomic impulses per unit of time may have permanently adjusted to a higher sensitivity than organs which receive many impulses per unit of time. As a corollary, the former may show very little postjunctional supersensitivity (but pronounced subsensitivity) on appropriate changes in the concentration of transmitters (or related agents) in the biophase, while the reverse may apply to organs with a heavy flow of nerve impulses.

V. Concluding Remarks

Although postjunctional supersensitivity has been thoroughly characterized in only a few tissues, the authors believe it is a nearly universal phenomenon and thus subscribe to the general ideas expressed in CANNON's "Law of Denervation" (CANNON and ROSENBLUETH, 1949). As far-sighted as that statement was, it may nevertheless be appropriate at this time to amend the "Law of Denervation" to the following:

When nerve function is chronically interrupted, be it by surgical, physiological, pathological or pharmacological means, most of the distal effectors become supersensitive to any process which initiates a response in the effector.

This statement differs from the original (Section I, p. 56) in several ways.

1. It is now known that the occurrence of supersensitivity is not restricted to situations in which the nerve has been surgically severed. Other procedures which interfere with normal nerve activity may initiate the phenomenon.

2. Because denervated ganglia do not exhibit supersensitivity, the statement has been modified to *most* of the distal effectors rather than to *all* of the distal effectors.

3. The supersensitive state of an effector is not necessarily restricted to chemical agents and neurotransmitters. For example, supersensitivity of smooth muscle to electrical stimulation has been observed.

4. In CANNON's "Law of Denervation" it was stated that "... supersensitivity is greater for the links which immediately follow the cut neurons and decreases progressively for more distal elements." This idea is absent from the present statement because it is a special circumstance related to conditions under which pre- and postjunctional supersensitivity coexist.

Postjunctional supersensitivity appears to be due to one or more of three mechanisms, depending upon the tissue considered. In skeletal muscle, a spread of receptors out from the end plate is the primary mechanism. Probable secondary mechanisms include changes in the electrical properties of the cell membrane and movement and/or binding of ions. In contrast, postjunctional supersensitivity in smooth and cardiac muscle probably depends primarily on changes in the electrophysiological properties of the cell membrane and of calcium binding and/or movement. There is no evidence to support a change in the number or properties of receptors in supersensitive smooth or cardiac muscle. A decision on the relative importance of these mechanisms in other tissues must await further research.

It is clear that subsensitivity is also an important aspect of sensitivity phenomena and the authors believe that postjunctional supersensitivity and postjunctional subsensitivity are probably opposite expressions of the same phenomenon. The chief deterrent to completely accepting this idea is the lack of quantitative data establishing the nonspecificity of subsensitive changes. This is partially due to the failure to find the optimum experimental conditions in which to study subsensitivity. If this obstacle is overcome and it is eventually established that the two types of sensitivity change are opposite expressions of the same phenomenon, the "Law of Denervation", should be modified to become the "Law of Innervation":

When functional nerve activity is chronically increased or decreased (surgically, physiologically, pathologically or pharmacologically) the sensitivity of most distal effectors to any process which initiates a response in the effector is slowly altered in a direction which will compensate for the altered neural input.

There are enough examples of supersensitivity and subsensitivity in man to indicate that these phenomena can create therapeutic complications. In the future, attention should be focused on the possibility of supersensitivity playing a significant part in drug interactions and the appearance of tolerance and/or withdrawal reactions to chronically used drugs such as antihypertensives and central nervous system stimulants and depressants.

The study of supersensitivity and subsensitivity is the study of the extremes of the normal sensitivity of tissues and organs. It is hoped that the information gained through such study will eventually aid in understanding the normal as well as the abnormal physiological response. We are at present a long way from achieving that understanding, but we have begun. This situation is similar to many others in basic research and can be best described by quoting Edith Bülbring who wrote in the introduction to the book *Smooth Muscle:* "It is not yet possible to assemble the multitide of observations and construct a complete picture of the whole subject. Although much has been observed, little is known. Yet the foundations have been laid and we can look forward to future experiments with confidence."

Acknowledgements. The authors gratefully acknowledge the benefit they derived from the *Symposium on Supersensitivity and Subsensitivity of Tissues to Drugs* held at West Virginia University, June 3–4, 1970. Thanks are due to the speakers at that symposium as well as to the organizations which generously sponsored it. The speakers were: Drs. Bito, Crout, de la Lande, Emmelin, Langer, Maxwell, Nickerson, Thesleff, Trendelenburg, Volle, Fleming and McPhillips. Dr. Sharpless, who was unable to attend, was greatly missed. The sponsors were: Astra Pharmaceutical Products, Inc., Burroughs-Wellcome & Co., Ciba Pharmaceutical Co., Hoffmann-La Roche, Inc., McNeil Laboratories, Mead-Johnson Research Center, West Virginia Heart Association and West Virginia University Foundation.

The authors also wish to express their appreciation for the support their research on supersensitivity and subsensitivity has received from the following grants: National Institutes of Health, U.S.P.H.S., grants number NB-03034, NS-08300, GM 2G-76, and ES-00396; West Virginia Heart Association grants number 66-AG-16-N, and 68-AG-4-N.

References

Abboud, F. M., Eckstein, J. W.: Effects of small oral doses of reserpine on vascular responses to tyramine and norepinephrine in man. Circulation **29**, 219–223 (1964a).

Abboud, F. M., Eckstein, J. W.: Venous and arterial responses to norepinephrine in dogs treated with reserpine. Amer. J. Physiol. **206**, 299–303 (1964b).

Adham, N., Schenk, E. A.: Autonomic innervation of the rat vagina, cervix and uterus and its cyclic variation. Amer. J. Obstet. Gynec. **104**, 508–516 (1969).

ALBUQUERQUE, E. X., THESLEFF, S.: A comparative study of membrane properties of innervated and chronically denervated fast and slow skeletal muscles in the rat. Acta physiol. scand. **73**, 471–480 (1968).

ALBUQUERQUE, E. X., WARNICK, J. E.: The pharmacology of batrachotoxin. IV. Interactions with tetradotoxin on innervated and chronically denervated rat skeletal muscle. J. Pharmacol. exp. Ther. **180**, 683–697 (1972).

ALONSO-DEFLORIDA, F., DEL CASTILLO, J., GONZALEZ, C. C., SANCHEZ, V.: Anaphylactic reaction of denervated skeletal muscle in the guinea pig. Science **147**, 1155–1156 (1965a).

ALONSO-DEFLORIDA, F., DEL CASTILLO, J., GONZALEZ, C. C., SANCHEZ, V.: On the pharmacological and anaphylactic responsiveness of denervated skeletal muscle of the guinea pig. Brit. J. Pharmacol. **25**, 610–620 (1965b).

BARNES, J. M., DENZ, F. A.: The reaction of rats to diets containing octamethyl pyrophosphoramide (Schraden) and 0,0-diethyl-S-ethylmercaptoethanol-thiophosphate ("Systox"). Brit. J. industr. Med. **11**, 11–19 (1954).

BAUER, H., GOODFORD, P. J., HÜTER, J.: The calcium content and calcium uptake of the smooth muscle of the guinea-pig taenia coli. J. Physiol. (Lond.) **176**, 163–179 (1965).

BIRMINGHAM, A. T., PATERSON, G., WÓJCICKI, J.: A comparison of the sensitivities of innervated and denervated rat vasa deferentia to agonist drugs. Brit. J. Pharmacol. **39**, 748–754 (1970).

BITO, L. Z., DAWSON, M. J.: The site and mechanism of the control of cholinergic sensitivity. J. Pharmacol. exp. Ther. **175**, 673–684 (1970).

BITO, L. Z., DAWSON, M. J., PETRINOVIC, L.: Cholinergic sensitivity: normal variability as a function of stimulus background. Science **172**, 583–585 (1971).

BITO, L. Z., HYSLOP, K., HYNDMAN, J.: Antiparasympathomimetic effects of cholinesterase inhibitor treatment. J. Pharmacol. exp. Ther. **157**, 159–169 (1967).

BOHR, D. F.: Electrolytes and smooth muscle contraction. Pharmacol. Rev. **16**, 85–111 (1964).

BOKRI, E., FEHÉR, O., MÒZSIK, GY.: Investigation of denervation supersensitivity in a sympathetic ganglion. Pflügers Arch. ges. Physiol. **277**, 347–356 (1963).

BOMBINSKI, T. J., DUBOIS, K. P.: Toxicity and mechanism of action of Di-Syston. Arch. Ind. Health **17**, 192–199 (1958).

BRODEUR, J., DUBOIS, K. P.: Studies on the mechanism of acquired tolerance by rats to 0,0-diethyl S-2 (ethylthio) ethyl phosphorodithioate (Di-Syston). Arch. int. Pharmacodyn. **149**, 560–570 (1964).

BRODY, I. A.: Relaxing factor in denervated muscle: a possible explanation for fibrillations. Amer. J. Physiol. **211**, 1277–1280 (1966).

BRODY, M. J.: Cardiovascular responses following immunological sympathectomy. Circulat. Res. **15**, 161–167 (1964).

BRODY, M. J., DIXON, R. L.: Vascular reactivity in experimental diabetes mellitus. Circulat. Res. **14**, 494–501 (1964).

BROWN, D. A.: Depolarization of normal and preganglionically denervated superior cervical ganglia by stimulant drugs. Brit. J. Pharmacol. **26**, 511–520 (1966).

BROWNLEE, G., JOHNSON, E. S.: The site of the 5-hydroxytryptamine receptor on the intramural plexus of the guinea-pig isolated ileum. Brit. J. Pharmacol. **21**, 306–322 (1963).

BROWNLEE, G., JOHNSON, E. S.: The release of acetylcholine from the isolated ileum of the guinea-pig induced by 5-hydroxytryptamine and dimethylphenylpiperazinium. Brit. J. Pharmacol. **24**, 689–700 (1965).

BUDGE, J. L.: Über die Bewegung der Iris: Für Physiologen und Ärzte, 206 S. Braunschweig: Vieweg 1855.

BURFORD, H. J., GILL, J. B.: Calcium secretion in normal and supersensitive submaxillary glands of the cat. Biochem. Pharmacol. **17**, 1881–1892 (1968).

BURN, J. H., RAND, M. J.: The cause of the supersensitivity of smooth muscle to noradrenaline after sympathetic degeneration. J. Physiol. (Lond.) **147**, 135–143 (1959).

BURN, J. H., RAND, M. J.: Noradrenaline in artery walls and its dispersal by reserpine. Brit. med. J. **1958 I**, 903–908.

BURNSTOCK, G.: Structure of smooth muscle and its innervation. In: Smooth muscle (ed. by: BÜLBRING, E.; BRADING, A. F.; JONES, A. W.; and TOMITA, T.). Baltimore: Williams & Wilkins 1970.

Cannon, W. B., Rosenblueth, A.: The supersensitivity of denervated structures. New York: Macmillan 1949.

Carmeliet, E., Vereecke, J.: Adrenaline and the plateau phase of the cardiac action potential. Importance of Ca^{++}, Na^+ and K^+ conductance. Pflügers Arch. ges. Physiol. **313**, 300–315 (1969).

Carrier, O., Douglas, B. H., Garrett, L., Whittington, P. J.: The effect of reserpine on vacsular tissue sodium and potassium content. J. Pharmacol. exp. Ther. **158**, 494–503 (1967).

Carrier, O., Jurevics, H. A.: The role of calcium in "nonspecific" supersensitivity of vascular muscle. J. Pharmacol. exp. Ther. (1972, in press).

Carrier, O., Shibata, S.: A possible role for tissue calcium in reserpine supersensitivity. J. Pharmacol. exp. Ther. **155**, 42–49 (1967).

Cervoni, P., Kirpekar, S. M.: Studies on the decentralized nictitating membrane of the cat. I. Effect of postganglionic electrical stimulation on the response to exogenous catecholamines. J. Pharmacol. exp. Ther. **152**, 8–17 (1966).

Cervoni, P., Reit, E., McCullough, J.: Studies on the decentralized nictitating membranes of the cat. II. Uptake and retention of norepinephrine and epinephrine. J. Pharmacol. exp. Ther. **175**, 649–663 (1970).

Chien, S.: Supersensitivity of denervated superior cervical ganglion to acetylcholine. Amer. J. Physiol. **198**, 949–954 (1960).

Clarke, D. E., Adams, H. R., Buckley, J. P.: Chronic reserpine treatment on adrenergic neuronal and receptor function in the isolated perfused mesenteric blood vessels of the dog. Europ. J. Pharmacol. **12**, 378–381 (1970).

Corey, S.: Changes in the sensitivity of the rat heart to drugs after chronic cholinesterase inhibition. Ph.D. dissertation, West Virginia University 1970.

Corey, S. E., McPhillips, J. J.: Supersensitivity to the negative chronotropic action of carbachol and methacholine in the rat. Brit. J. Pharmacol. **44**, 586–588 (1972).

Creese, R., El-Shafie, A. L., Vrbová, G.: Sodium movements in denervated muscle and the effects of antimycin A. J. Physiol. (Lond.) **197**, 279–294 (1968).

Creese, R., Taylor, D. B., Case, R.: Labeled decamethonium in denervated skeletal muscle. J. Pharmacol. exp. Ther. **176**, 418–422 (1971).

Crout, J. R., Muskus, A. J., Trendelenburg, U.: Effect of tyramine on isolated guinea-pig atria in relation to their noradrenaline stores. Brit. J. Pharmacol. **18**, 600–611 (1962).

Dahlström, A., Häggendal, J.: Recovery of noradrenaline levels after reserpine compared with the life span of amine storage granules in rat and rabbit. J. Pharm. Pharmacol. **18**, 750–752 (1966).

Dawkins, O., Bohr, D. F.: Sodium and potassium movement in the excised rat aorta. Amer. J. Physiol. **199**, 28–30 (1960).

Day, M., Vane, J. R.: An analysis of the direct and indirect actions of drugs on the isolated guinea-pig ileum. Brit. J. Pharmacol. **20**, 150–170 (1963).

de la Lande, I. S., Frewin, D., Waterson, J., Canell, V.: Factors influencing supersensitivity to noradrenaline in the isolated perfused artery; comparative effects of cocaine, denervation and serotonin. Circulat. Res. **20–21**, Suppl. III, 177–181 (1967).

de Moraes, S., Carvalho, F. V., Wehrle, R. D.: Sensitivity changes to noradrenaline in the guinea-pig vas deferens induced by amphetamine, cocaine and denervation. J. Pharm. Pharmacol. **22**, 717–719 (1970).

Dempsey, P. J., Cooper, T.: Supersensitivity of the chronically denervated feline heart. Amer. J. Physiol. **215**, 1245–1249 (1968).

Dominic, J. A., Moore, K. E.: Supersensitivity to the central stimulant actions of adrenergic drugs following discontinuation of a chronic diet of α-methyltyrosine. Psychopharmacologia (Berl.) **15**, 96–101 (1969).

Emmelin, N.: Supersensitivity following "pharmacological denervation". Pharmacol. Rev. **13**, 17–37 (1961).

Emmelin, N.: Submaxillary and sublingual secretion in cats during degeneration of post-ganglionic parasympathetic fibers. J. Physiol. (Lond.) **162**, 270–281 (1962).

Emmelin, N.: Action of acetylcholine on the responsiveness of effector cells. Experientia (Basel) **20**, 275 (1964).

Emmelin, N.: Action of transmitters on the responsiveness of effector cells. Experientia (Basel) **21**, 57–65 (1965).

EMMELIN, N., STRÖMBLAD, B. C. R.: A "paroxysmal" secretion of saliva following parasympathetic denervation of the parotid gland. J. Physiol. (Lond.) 143, 506–514 (1958).

EVANS, D. H. L., SCHILD, H. O., THESLEFF, S.: Effects of drugs on depolarized plain muscle. J. Physiol. (Lond.) 143, 474–485 (1958).

FLEMING, W. W.: Supersensitivity of the cat heart to catecholamine-induced arrhythmias following reserpine pre-treatment. Proc. Soc. exp. Biol. (N.Y.) 111, 484–486 (1962).

FLEMING, W. W.: A comparative study of supersensitivity to norepinephrine and acetylcholine produced by denervation, decentralization and reserpine. J. Pharmacol. exp. Ther. 141, 173–179 (1963a).

FLEMING, W. W.: Changes in the sensitivity of the cat's nictitating membrane to norepinephrine, acetylcholine and potassium. Biochem. Pharmacol. 12 (Suppl.), 202 (1963b).

FLEMING, W. W.: Nonspecific supersensitivity of the guinea-pig ileum produced by chronic ganglion blockade. J. Pharmacol. exp. Ther. 162, 277–285 (1968).

FLEMING, W. W.: Supersensitivity of the denervated rat diaphragm to potassium:, A comparison with supersensitivity in other tissues. J. Pharmacol. exp. Ther. 176, 160–166 (1971).

FLEMING, W. W.: Altered resting membrane potential (RMP) of supersensitive smooth muscle cells. Abstratsc of the Fifth International Congr. on Pharmacology (1972, in press).

FLEMING, W. W., TRENDELENBURG, U.: The development of supersensitivity to norepinephrine after pretreatment with reserpine. J. Pharmacol. exp. Ther. 133, 41–51 (1961).

FLEMING, W. W., WESTFALL, D. P., DE LA LANDE, I. S., JELLETT, L. B.: Log-normal distribution of equieffective doses of norepinephrine and acetylcholine in several tissues. J. Pharmacol. exp. Ther. 181, 339–345 (1972).

FOLEY, D. J., MCPHILLIPS, J. J.: Sensitivity of the rat uterus and ileum to carbachol following subacute administration of a cholinesterase inhibitor. Fed. Proc. 30, 621 (1971).

FOSTER, R. W.: The nature of the adrenergic receptors of the trachea of the guinea-pig. J. Pharm. Pharmacol. 18, 1–12 (1966).

FREWIN, D. B., GILMORE, H. R., HO, J. Q. K., SCROOP, G. C.: Clinical, physiological and pathological observations in a case of progressive autonomic nervous system degeneration associated with Holmes-Adie Syndrome and peripheral neuropathy. Aust. Ann. Med. 17, 141–147 (1968).

FRIEDMAN, M. J., JAFFE, J. H., SHARPLESS, S. K.: Central nervous system supersensitivity to pilocarpine after withdrawal of chronically administered scopolamine. J. Pharmacol. exp. Ther. 167, 45–55 (1969).

GARRETT, R. L., CARRIER, O.: Alteration of extracellular calcium dependence in vascular tissue by reserpine. Europ. J. Pharmacol. 13, 306–311 (1971).

GILLIS, C. N.: The effect of acute and chronic administration of reserpine on the respiration of rat aorta. J. Pharmacol. exp. Ther. 127, 265–267 (1959).

GOODMAN, F. R., WEISS, G. B.: Effects of lanthanum on ^{45}Ca movements and on contractions induced by norepinephrine, histamine and potassium in vascular smooth muscle. J. Pharmacol. exp. Ther. 177, 415–425 (1971).

GRAEFE, K. H., BÖNISCH, H., TRENDELENBURG, U.: Time-dependent changes in neuronal net uptake of noradrenaline after pretreatment with pargyline and/or reserpine. Naunyn-Schmiedebergs Arch. Pharmak. 271, 1–28 (1971).

GREEN, A. F., ROBSON, R. D.: Adrenergic neurone blocking agents: tolerance and hypersensitivity to adrenaline and noradrenaline. Brit. J. Pharmacol. 25, 497–506 (1965).

GREEN, R. D.: The effect of denervation on the sensitivity of the superior cervical ganglion of the pithed cat. J. Pharmacol. exp. Ther. 167, 143–150 (1969).

GREEN, R. D., FLEMING, W. W.: Agonist-antagonist interactions in the normal and supersensitive nictitating membrane of the spinal cat. J. Pharmacol. exp. Ther. 156, 207–214 (1967).

GREEN, R. D., FLEMING, W. W.: Analysis of supersensitivity in the isolated spleen of the cat. J. Pharmacol. exp. Ther. 162, 254–262 (1968).

GREEN, R. D., FLEMING, W. W., SCHMIDT, J. L.: Sensitivity changes in the isolated ileum of the guinea pig after pretreatment with reserpine. J. Pharmacol. exp. Ther. 162, 270–276 (1968).

GUTMANN, E., SANDOW, A.: Caffeine-induced contracture and potentiation of contraction in normal and denervated rat muscle. Life Sci. **4**, 1149–1156 (1965).

HAEUSLER, G., HAEFELY, W.: Pre- and postjunctional supersensitivity of the mesenteric artery preparation from normotensive and hypertensive rats. Naunyn-Schmiedebergs Arch. Pharmak. **266**, 18–33 (1970).

HAEUSLER, G., HAEFELY, W., THOENEN, H.: Chemical sympathectomy of the cat with 6-hydroxytryptamine. J. Pharmacol. exp. Ther. **170**, 50–61 (1969).

HAMPLE, C. W.: The effect of initial tension and load on the response of the nictitating membrane of the cat. Amer. J. Physiol. **107**, 717–725 (1934).

HINKE, J. A. M.: Calcium requirements for noradrenaline and high potassium ion contraction in arterial smooth muscle. In: Muscle (ed. by: PAUL, W. M.; DANIEL, E. E.; KAY, C. M.; and MONEKTON, G. J.). New York: Pergamon Press 1965.

HUBBARD, S. J.: The electrical constants and the component conductances of frog skeletal muscle after denervation. J. Physiol. (Lond.) **165**, 443–456 (1963).

HUDGINS, P. M., FLEMING, W. W.: A relatively nonspecific supersensitivity in aortic strips resulting from pretreatment with reserpine. J. Pharmacol. exp. Ther. **153**, 70–80 (1966).

HUDGINS, P. M., HARRIS, T. M.: Further studies on the effects of reserpine pretreatment on rabbit aorta: calcium and histologic changes. J. Pharmacol. exp. Ther. **175**, 609–618 (1970).

HUDGINS, P. M., WEISS, G. B.: Differential effects of calcium removal upon vascular smooth muscle contraction induced by norepinephrine, histamine and potassium. J. Pharmacol. exp. Ther. **159**, 91–97 (1968).

INOMATA, H., SUZUKI, T.: Cholinergic spontaneous junction potentials in guinea-pig vas deferens. Tohoku J. exp. Med. **105**, 197–198 (1971).

ISAACSON, A., SANDOW, A.: Caffeine effects on radiocalcium movement in normal and denervated rat skeletal muscle. J. Pharmacol. exp. Ther. **155**, 376–388 (1967).

IWAYAMA, T., FLEMING, W. W., BURNSTOCK, G.: Ultrastructure of mitochondria in atrial muscle associated with depression and supersensitivity produced by reserpine. J. Pharmacol. exp. Ther. **184**, 95–105 (1973).

JOHANSSON, B., LJUNG, B., MALMFORS, T., OLSON, L.: Prejunctional supersensitivity in the rat portal vein as related to its pattern of innervation. Acta physiol. scand., Suppl. **349**, 5–16 (1970).

JOHNS, R. J., McQUILLEN, M. P.: Syndromes simulating myasthenia gravis: asthenia with anticholinesterase tolerance. Ann. N.Y. Acad. Sci. **135**, 385–397 (1966).

JONES, A.: Ganglionic actions of muscarine substances. J. Pharmacol. exp. Ther. **141**, 195–205 (1963).

KALSNER, S., NICKERSON, M.: Disposition of norepinephrine and epinephrine in vascular tissue, determined by the technique of oil immersion. J. Pharmacol. exp. Ther. **165**, 152–165 (1969).

KASUYA, Y., GOTO, K.: The mechanism of supersensitivity to norepinephrine induced by cocaine in rat isolated vas deferens. Europ. J. Pharmacol. **4**, 355–362 (1968).

KASUYA, K., GOTO, K., HASHIMOTO, H., WATANABE, H., MUNAKATA, H., WATANABE, M.: Nonspecific denervation supersensitivity in the rat vas deferens *in vitro*. Europ. J. Pharmacol. **8**, 177–184 (1969).

KIRPEKAR, S. M., CERVONI, P., FURCHGOTT, R. F.: Catecholamine content of the cat nictitating membrane following procedures sensitizing it to norepinephrine. J. Pharmacol. exp. Ther. **135**, 180–190 (1962).

KLAUS, W., LÜLLMANN, H., MUSCHOLL, E.: Der Kalium-Flux des normalen und denervierten Rattenzwerchfells. Pflügers Arch. ges. Physiol. **271**, 761–775 (1960).

KOCH-WESER, J., BLINKS, J. R.: The influence of the interval between beats on myocardial contractility. Pharmacol. Rev. **15**, 601–652 (1963).

KOVACIC, B., ROBINSON, R. L.: The effect of reserpine on catecholamine levels in the gravid rat and its offspring. J. Pharmacol. exp. Ther. **152**, 37–41 (1966).

KURIYAMA, H.: Effects of ions and drugs on the electrical activity of smooth muscle. In: Smooth muscle (ed. by: BÜLBRING, E.; BRADING, A. F.; JONES, A. W.; and TOMITA, T.). Baltimore: Williams & Wilkins 1970.

LANGENDORFF, O.: Die Deutung der „paradoxen" Pupillenerweiterung. Klin. Mbl. Augenheilk. **38**, 823–827 (1900).

LANGER, S. Z.: Presence of tone in the denervated and in the decentralized nictitating membrane of the spinal cat and its influence on determinations of supersensitivity. J. Pharmacol. exp. Ther. **154**, 14–34 (1966).

LANGER, S. Z., DRASKÒCZY, P. R., TRENDELENBURG, U.: Time course of the development of supersensitivity to various amines in the nictitating membrane of the pithed cat after denervation or decentralization. J. Pharmacol. exp. Ther. **157**, 255–273 (1967).

LANGER, S. Z.: TRENDELENBURG, U.: Studies on veratrum alkaloids: interaction of veratramine and accelerating agents on the heart. J. Pharmacol. exp. Ther. **146**, 99–110 (1964).

LANGER, S. Z., TRENDELENBURG, U.: Decreased in effectiveness of phenoxybenzamine after chronic denervation and chronic decentralization of the nictitating membrane of the pithed cat. J. Pharmacol. exp. Ther. **163**, 290–299 (1968).

LANGER, S. Z., TRENDELENBURG, U.: The effect of a saturable uptake mechanism on the slopes of dose-response curves for sympathomimetic amines and on the shifts of dose-response curves produced by a competitive antagonist. J. Pharmacol. exp. Ther. **167**, 117–142 (1969).

LANGHAM, M. E., FRASER, L. K.: The absence of supersensitivity to adrenergic amines in the eye of the conscious rabbit following preganglionic cervical sympathectomy. Life Sci. **5**, 1699–1705 (1966).

LEE, F. L.: The relation between norepinephrine content and response to sympathetic nerve stimulation of various organs of cats pretreated with reserpine. J. Pharmacol. exp. Ther. **156**, 137–141 (1967).

LENMAN, J. A. R.: Effect of denervation on the resting membrane potential of healthy and dystrophic muscle. J. Neurol. Neurosurg. Psychiat. **28**, 525–528 (1965).

LI, C.-L.: Mechanism of fibrillation potentials in denervated mammalian skeletal muscle. Science **132**, 1889–1890 (1960).

LOWN, B., EHRLICH, L., LIPSCHULTZ, B., BLAKE, J.: Effect of digitalis in patients receiving reserpine. Circulation **24**, 1185–1191 (1961).

MAGLADERY, J. W., SOLANDT, D. Y.: Relation of fibrillation to acetylcholine and potassium sensitivity in denervated skeletal muscle. J. Neurophysiol. **5**, 357–362 (1942).

MALING, H. M., FLEISCH, J. H., SAUL, W. F.: Species differences in aortic response to vasoactive amines: the effects of compound 48/80, cocaine, reserpine and 6-hydroxytryptamine. J. Pharmacol. exp. Ther. **176**, 672–683 (1971).

MAXWELL, R. A., PLUMMER, A. J., POVALSKI, H., SCHNEIDER, F.: Concerning a possible action of guanethidine (Su-5864) in smooth muscle. J. Pharmacol. exp. Ther. **129**, 24–30 (1960).

MAXWELL, R. A., WASTILA, W. B., ECKHARDT, S. B.: Some factors determining the response of rabbit aortic strips to dl-norepinephrine-7-H^3 hydrochloride and the influence of cocaine, guanethidine and methylphenidate on these factors. J. Pharmacol. exp. Ther. **151**, 253–261 (1966).

McCLURE, D. C., WESTFALL, D. P., FLEMING, W. W.: The effect of postganglionic denervation on the sensitivity of the smooth muscle of the guinea pig vas deferens to electrical stimulation. Pharmacologist **13**, 199 (1971).

McNEILL, J. H.: Reserpine supersensitivity to catecholamine-induced cardiac phosphorylase activation. Canad. J. Physiol. Pharmacol. **47**, 515–519 (1969).

McNEILL, J. H.: Potentiation of norepinephrine-induced activation of cardiac phosphorylase by theophylline and reserpine. Canad. J. Physiol. Pharmacol. **48**, 149–151 (1970).

McNEILL, J. H., SCHULZE, S.: Reserpine-induced supersensitivity to histamine activated cardiac phosphorylase and cardiac contractility. Res. Comm. Chem. Path. Pharmacol. **3**, 339–347 (1972).

McPHILLIPS, J. J.: Subsensitivity of the rat ileum to cholinergic drugs. J. Pharmacol. exp. Ther. **166**, 249–254 (1969).

McPHILLIPS, J. J., DAR, M. S.: Resistance to the effects of carbachol on the cardiovascular system and on the isolated ileum of rats after subacute administration of an organophosphorus cholinesterase inhibitor. J. Pharmacol. exp. Ther. **156**, 507–513 (1967).

MILEDI, R.: Induction of receptors. In: Ciba Foundation Symposium on Enzymes and Drug Action (ed. by: MONGAR, L. J.; and DE REUCK, A. V. S.). Boston: Little, Brown and Co. 1962.

Miledi, R.: An influence of nerve not mediated by impulses. In: The effect of use and disuse on neuromuscular functions (ed. by: Gutmann, E.; and Hnik, P.). Amsterdam-London-New York: Elsevier 1963.

Morrison, J. M., Fleming, W. W.: Supersensitivity of decentralized and denervated nictitating membranes to barium. Proc. Soc. exp. Biol. (N.Y.) 136, 196–199 (1971).

Mózsik, G., Jávor, T., Dobi, S., Petrássy, K., Szabó, A.: Development of "pharmacological denervation phenomenon" in patients treated with atropine. Europ. J. Pharmacol. 1, 391–395 (1967).

Muelheims, G. H., Entrup, R. W., Paiewonsky, D., Mierzwiak, D. S.: Increased sensitivity of the heart to catecholamine-induced arrhythmias following guanethidine. Clin. Pharmacol. Ther. 6, 757–762 (1965).

Oliver, W. T., Funnell, H. S.: Correlation of effects of parathion on erythrocyte cholinesterase with symptomatology in pigs. Amer. J. vet. Res. 22, 80–84 (1961).

Ord, M. G., Thompson, R. H. S.: The distribution of cholinesterase types in mammalian tissues. Biochem. J. 46, 346–352 (1950).

Orlans, F. B. H., Finger, K. F., Brodie, B. B.: Pharmacological consequences of the selective release of peripheral norepinephrine by syrosingopine (SU-3118). J. Pharmacol. exp. Ther. 128, 131–139 (1960).

Ozawa, H., Sugawara, K.: Sensitivity of the isolated vas deferens of the guinea-pig to norepinephrine and acetylcholine after denervation, decentralization and treatment by various agents. Europ. J. Pharmacol. 11, 56–66 (1970).

Pellegrino, C., Franzini, C.: An electron microscope study of denervation atrophy in red and white skeletal muscle fibers. J. Cell Biol. 17, 327–349 (1963).

Perrine, S. E., McPhillips, J. J.: Specific subsensitivity of the rat atrium to cholinergic drugs. J. Pharmacol. exp. Ther. 175, 496–502 (1970).

Philipeaux, J. M., Vulpian, A.: Note sur une modification physiologique qui se produit dans le nerf lingual par suite de l'abolition temporaire de la motricité dans le nerf hypoglosse du même côté. C. R. Acad. Sci. (Paris) 56, 1009–1011 (1863).

Pluchino, S.: Direct and indirect effects of 5-hydroxytryptamine and tyramine on cat smooth muscle. Naunyn-Schmiedebergs 'Arch. Pharmacol. 272, 189–224 (1972).

Pluchino, S., Trendelenburg, U.: The influence of denervation and of decentralization on the alpha and beta effects of isoproterenol on the nictitating membrane of the pithed cat. J. Pharmacol. exp. Ther. 163, 257–265 (1968).

Reas, H. W., Trendelenburg, U.: Changes in the sensitivity of the sweat glands of the cat after denervation. J. Pharmacol. exp. Ther. 156, 126–136 (1967).

Redfern, P., Lundh, H., Thesleff, S.: Tetrodotoxin resistant action potentials in denervated rat skeletal muscle. Europ. J. Pharmacol. 11, 263–265 (1970).

Redfern, P., Thesleff, S.: Action potential generation in denervated rat skeletal muscle. I. Quantitative aspects. Acta physiol. scand. 81, 557–564 (1971).

Rider, J. A., Ellinwood, L. E., Coon, J. M.: Production of tolerance in the rat to octamethyl pyrophosphoramide (OMPA). Proc. Soc. exp. Biol. (N.Y.) 81, 455–459 (1952).

Romano, D. V., Cervoni, P., McGrath, W. R.: Effects of decentralization on the in vitro responses of the guinea pig vas deferens. Pharmacologist 13, 200 (1971).

Rothballer, A. B., Sharpless, S. K.: Effects of intracranial stimulation on denervated nictitating membrane of the cat. Amer. J. Physiol. 200, 901–908 (1961).

Sakurai, T., Hashimoto, Y.: The vasoconstrictor action of angiotensin in relation to catecholamines. Jap. J. Pharmacol. 15, 223–233 (1965).

Schaeppi, U.: Comparison of serotonin with directly and indirectly acting compounds in their effect on the nictitating membrane of spinal cats. J. Pharmacol. exp. Ther. 139, 330–336 (1963).

Schaeppi, U.: Postganglionic nature of parasympathetic innervation of pig iris sphincter. Amer. J. Physiol. 210, 91–94 (1966).

Schmidt, J. L., Fleming, W. W.: The structure of sympathomimetics as related to reserpine induced sensitivity changes in the rabbit ileum. J. Pharmacol. exp. Ther. 139, 230–237 (1963).

Schmidt, J. L., Fleming, W. W.: A nonsympathomimetic effect of cyclopentamine and β-mercaptoethylamine in the rabbit ileum. J. Pharmacol. exp. Ther. 145, 83–86 (1964).

Schneyer, L. H., Yoshida, Y.: Secretory potentials in rat submaxillary gland. Proc. Soc. exp. Biol. (N.Y.) 130, 192–196 (1969).

SCHWARTZ, A., LEE, K. S.: Effect of reserpine on heart mitochondria. Nature (Lond.) **188**, 948–949 (1960).

SEIDEHAMEL, R. J., PATIL, P. N., TYE, A., LAPIDUS, J. B.: The effects of norepinephrine isomers on various supersensitivities of the cat nictitating membranes. J. Pharmacol. exp. Ther. **153**, 81–89 (1966).

SHARPLESS, S. K.: Reorganization of function in the nervous system—use and disuse. Ann. Rev. Physiol. **26**, 357–388 (1964).

SHARPLESS, S. K.: Isolated and deafferented neurons: disuse supersensitivity. In: Basic mechanisms of the epilepsies (ed. by: JAPSER, H. H.; WARD, A. A.; and POPE, A.). Boston: Little, Brown 1969.

SHIBATA, S., KUCHII, M., KURAHASHI, K.: The supersensitivity of isolated rabbit atria and aortic strips produced by 6-hydroxydopamine. Europ. J. Pharmacol. **18**, 271–280 (1972).

SJÖSTRAND, N. O., SWEDIN, G.: Effect of reserpine on the noradrenaline content of the vas deferens and the seminal vesicle compared with the submaxillary gland and the heart of the rat. Acta physiol. scand. **72**, 370–377 (1968).

SMITH, A. A., DANCIS, J.: Physiologic studies in familial dysautonomia. J. Pediat. **63**, 838–840 (1963).

SMITH, C. B.: Relaxation of the nictitating membrane of the spinal cat by sympathomimetic amines. J. Pharmacol. exp. Ther. **142**, 163–170 (1963).

SOMLYO, A. P., SOMLYO, A. V.: Vascular smooth muscle. I. Normal structure, pathology, biochemistry and biophysics. Pharmacol. Rev. **20**, 197–272 (1968).

SPEHLMANN, R.: Acetylcholine and the epilepti-form activity of chronically isolated cortex. Arch. Neurol. **24**, 495–502 (1971).

STAVRAKY, G. W.: Supersensitivity following lesions of the nervous system. Toronto: University of Toronto Press 1961.

SUN, S.-C., SOHOL, R. S., COLCOLOUGH, H. L., BURCH, G. E.: Histochemical and electron microscopic studies of the effects of reserpine on the heart muscle of mice. J. Pharmacol. exp. Ther. **161**, 210–221 (1968).

TAYLOR, J., GREEN, R. D.: Analysis of reserpine-induced supersensitivity in aortic strips of rabbits. J. Pharmacol. exp. Ther. **177**, 127–135 (1971).

TAYLOR, P. W., RICHARDSON, K. C., RODDY, P. M., TITUS, E.: A new effect of reserpine: accumulation of glycoprotein in the submaxillary gland. J. Pharmacol. exp. Ther. **156**, 483–491 (1967).

THESLEFF, S.: Effects of motor innervation on the chemical sensitivity of skeletal muscle. Physiol. Rev. **40**, 734–752 (1960).

THESLEFF, S.: Spontaneous electrical activity in denervated rat skeletal muscle. In: The effect of use and disuse on neuromuscular functions (ed. by: GUTMANN, E.; and HONIK, P.). Amsterdam-London-New York: Elsevier 1963.

TOMITA, T.: Electrical properties of mammalian smooth muscle. In: Smooth muscle (ed. by: BÜLBRING, E.; BRADING, A. F.; JONES, A. W.; and TOMITA, T.). Baltimore: Williams & Wilkins 1970.

TRENDELENBURG, U.: The action of 5-hydroxytryptamine on the nictitating membrane and on the superior cervical ganglion of the cat. Brit. J. Pharmacol. **11**, 74–80 (1956).

TRENDELENBURG, U.: Supersensitivity and subsensitivity to sympathomimetic amines. Pharmacol. Rev. **15**, 225–276 (1963 a).

TRENDELENBURG, U.: Time course of changes in sensitivity after denervation of the nictitating membrane of the spinal cat. J. Pharmacol. exp. Ther. **142**, 335–342 (1963 b).

TRENDELENBURG, U.: Supersensitivity by cocaine to dextrorotatory isomers of norepinephrine and epinephrine. J. Pharmacol. exp. Ther. **148**, 329–338 (1965).

TRENDELENBURG, U.: Mechanisms of supersensitivity and subsensitivity to sympathomimetic amines. Pharmacol. Rev. **18**, 629–640 (1966).

TRENDELENBURG, U.: Supersensitivity of the isolated nictitating membrane of the cat to sympathomimetic amines after impairment of the intraneuronal mechanisms of inactivation. Naunyn-Schmiedebergs Arch. Pharmak. **271**, 29–58 (1971).

TRENDELENBURG, U., DRASKÓCZY, P. R., PLUCHINO, S.: The density of adrenergic innervation of the cat's nictitating membrane as a factor influencing the sensitivity of the isolated preparation to 1-norepinephrine. J. Pharmacol. exp. Ther. **166**, 14–25 (1969).

TRENDELENBURG, U., MAXWELL, R. A., PLUCHINO, S.: Methoxamine as a tool to assess the importance of intraneuronal uptake of 1-norepinephrine in the cat's nictitating membrane. J. Pharmacol. exp. Ther. **172**, 91–99 (1970).

TRENDELENBURG, U., WEINER, N.: Sensitivity of the nictitating membrane after various procedures and agents. J. Pharmacol. exp. Ther. **136**, 152–161 (1962).

TSAI, T. H.: Sensitivity of the nictitating membrane (NM) of the pithed cat to infusion of 1-norepinephrine (NE) after various procedures and agents. Fed. Proc. **29**, 614 (1970).

TSAI, T. H., DENHAM, S., McGRATH, W. R.: Sensitivity of the isolated nictitating membrane of the cat to norepinephrine and acetylcholine after various procedures and agents. J. Pharmacol. exp. Ther. **164**, 146–157 (1968).

TSAI, T. H., PENN, J. T.: The response of isolated rat vas deferens to norepinephrine and acetylcholine after decentralization. Fed. Proc. **31**, 510 (1972).

URQUILLA, P. R., STITZEL, R. E., FLEMING, W. W.: The antagonism of phentolamine against exogenously administered and endogenously released norepinephrine in rabbit aortic strips. J. Pharmacol. exp. Ther. **172**, 310–319 (1970).

VAN ZWIETEN, P. A., WIDHALM, S., HERTTING, G.: Influence of cocaine and of pretreatment with reserpine on the pressor effect and the tissue uptake of injected dl-catecholamines-2-H^3. J. Pharmacol. exp. Ther. **149**, 50–56 (1965).

VARMA, D. R.: Effect of sympathetic denervation on the *alpha* receptors of the cat nictitating membrane. J. Pharmacol. exp. Ther. **153**, 48–61 (1966).

VICKERSON, F. H. L., VARMA, D. R.: Effects of denervation on the sensitivity of the superior cervical ganglion of the cat to acetylcholine and McN-A-343. Canad. J. Physiol. Pharmacol. **47**, 255–259 (1969).

VOLLE, R. L.: Modification by drugs of synaptic mechanisms in autonomic ganglia. Pharmacol. Rev. **18**, 839–870 (1966).

VOLLE, R. L., KOELLE, G. B.: The physiological role of acetylcholinesterase (ACHE) in sympathetic ganglia. J. Pharmacol. exp. Ther. **133**, 222–240 (1961).

WAGNER, K., TRENDELENBURG, U.: Development of degeneration contraction and supersensitivity in the cat's nictitating membrane after 6-hydroxydopamine. Naunyn-Schmiedebergs Arch. Pharmak. **270**, 215–236 (1971).

WAKADE, A. R., KIRPEKAR, S. M.: Chemical and histochemical studies on the sympathetic innervation of the vas deferens and seminal vesicle of the guinea pig. J. Pharmacol. exp. Ther. **178**, 432–441 (1971).

WAUD, D. R., KRAYER, O.: The rate-increasing effect of epinephrine and norepinephrine and its modification by experimental time in the isolated heart of normal and reserpine-pretreated dogs. J. Pharmacol. exp. Ther. **128**, 352–357 (1960).

WEISS, B.: Effects of environmental lighting and chronic denervation on the activation of adenyl cyclase of rat pineal gland by norepinephrine and sodium fluoride. J. Pharmacol. exp. Ther. **168**, 146–152 (1969).

WEISS, B., COSTA, E.: Adenylcyclase activity in rat pineal gland: effects of chronic denervation and norepinephrine. Science **156**, 1750–1751 (1967).

WESTFALL, D. P.: Nonspecific supersensitivity of the guinea-pig vas deferens produced by decentralization and reserpine treatment. Brit. J. Pharmacol. **39**, 110–120 (1970a).

WESTFALL, D. P.: The effect of reserpine treatment and decentralization on the ion distribution in the vas deferens of the guinea pig. Brit. J. Pharmacol. **39**, 121–127 (1970b).

WESTFALL, D. P., FLEMING, W. W.: The sensitivity of the guinea-pig pacemaker to norepinephrine and calcium after pretreatment with reserpine. J. Pharmacol. exp. Ther. **164**, 259–269 (1968a).

WESTFALL, D. P., FLEMING, W. W.: Sensitivity changes in the dog heart to norepinephrine, calcium and aminophylline resulting from pretreatment with reserpine. J. Pharmacol. exp. Ther. **159**, 98–106 (1968b).

WESTFALL, D. P., FLEMING, W. W.: Reserpine-induced supersensitivity in perfused rabbit hearts. Pharmacologist **10**, 218 (1968c).

WESTFALL, D. P., GILBERT, P. E., FLEMING, W. W.: Tension-response relationships in the intact nictitating membrane of the pithed cat. J. Pharmacol. exp. Ther. **169**, 196–200 (1969).

WESTFALL, D. P., McCLURE, D. C., FLEMING, W. W.: The effects of denervation, decentralization and cocaine on the response of the smooth muscle of the guinea-pig vas deferens to various drugs. J. Pharmacol. exp. Ther. **181**, 328–338 (1972).

WESTFALL, T. C., PEACH, M. J.: Influence of equilibration perfusion duration on H³-norepinephrine uptake and intracellular cation concentrations in isolated guinea-pig hearts. Pharmacologist 12, 234 (1970).

WILCKEN, D. E. L., BRENDER, D., MacDONALD, G. J., SHOREY, C. D., HINTERBERGER, H.: Effect of reserpine on the structure of heart mitochondria and the relation to catecholamine depletion. Circulat. Res. 21 (Suppl. III), 203–211 (1967).

WITHRINGTON, P., ZAIMIS, E.: The reserpine-treated cat. Brit. J. Pharmacol. 17, 380–391 (1961).

WITHRINGTON, P., ZAIMIS, E.: Cardiovascular effects produced in cats by the chronic administration of small doses of reserpine. Cardiovasc. Res. 1, 52–62 (1967).

ZAIMIS, E.: Reserpine-induced circulatory failure. Nature (Lond.) 192, 521–523 (1961).

ŽUPANČIČ, A. O.: Evidence for the identity of anionic centers of cholinesterase with cholinoreceptors. Ann. N.Y. Acad. Sci. 144, 689–693 (1967).

The Cell Surface in Cell Interactions

R. S. Turner** and M. M. Burger*

With 11 Figures

Table of Contents

The evolution of multicellular organisms requires the existence of controlled cell interactions. Such interactions can be divided into two main categories: those involved in cellular recognition and those which regulate cellular activity.

Cellular recognition is a fundamental process governing various functions present in multicellular organisms. Histogenesis during development and loss of organized structure during carcinogenesis are manifestations of the acquisition and loss of the ability of cells to recognize each other as a part of an organized multicellular unit.

The regulation of cellular activity produces the coordinated function of a variety of cell types which is a prerequisite for the evolution of multicellular organisms. Many investigators subscribe to the theory that carcinogenesis involves the breakdown of the cellular interactions which normally control cell growth.

The experimental analysis of cell interaction, whether concerned with cellular recognition or regulation of cellular activity, has as its goal the elucidation of the molecular basis of these phenomena. A number of approaches and biological systems have been used. Table 1 provides a comprehensive list of the cell interactions currently being investigated.

This review will discuss two types of cell interactions which are widely studied as model systems: first the reaggregation of dissociated cells *in vitro*,

* Department of Biochemistry, Biocenter of the University of Basel, Klingelbergstraße 70, CH-4056 Basel.
** R. S. Turner is a fellow of the Damon Runyon Memorial Fund for Cancer Research, Inc. Work conducted in the author's laboratory was supported by a grant and a contract from the National Cancer Institute, United States Public Health Service.

Table 1. Cell interactions

System	References	
Fertilization	Colwin and Colwin [49]	Dan [61]
	Metz and Monroy [149]	Gwatkin [96]
Positional information	Wolpert [247, 248, 249]	Crick [52, 53]
	Cohen [48]	Lawrence [132]
Embryonic induction	Weiss [239]	Grobstein [89]
	Zwilling [252]	Saunders [190]
Neuronal development	Weiss [240, 241, 242]	Sidman [213]
	Hamburger [95]	Bodian [14]
Nerve regeneration	Sperry [214, 10]	Gaze [82, 83, 84, 221]
	Jacobson [114, 115, 122]	
Immune response	Burnet [42]	Mitchison [149, 150]
	Nossal [170]	Raff [181]

in particular, the reaggregation of dissociated sponge cells, and secondly density-dependent inhibition of growth of cells in culture. The first system is a useful model for the analysis of specific cell recognition and the second is an example of cell interactions which are involved in the regulation of cellular activity.

Tissue Specific Reaggregation

The reaggregation of dissociated cells has been used in studies of cellular recognition since H. V. Wilson's [243] observation that a complete sponge could be reconstituted from a suspension of single cells.

Moscona [154, 160] instituted the use of proteolytic enzymes and chelators to dissociate vertebrate tissues and first conducted reaggregations in gently swirling flasks [155, 156]. His results indicate that reaggregation occurs in two stages: initial nonspecific contacts followed by sorting out in a histiotypic fashion, resulting in suspensions of cells obtained from more than one tissue forming aggregates in which the cells of the different tissues are segregated from each other (Moscona [153, 155, 157, 158, 202, 203]). When the same experiment is conducted with cells from the same tissue from two different species, the cells do not sort out, indicating that histiotypic specificity is expressed to a greater degree than species specificity [154, 157, 158, 161]. When sorting out does occur, it results in one cell type coming to lie external to the second cell type. Steinberg [216] has demonstrated that the internal-external relationship between any two given cell types is constant. The results of sorting-out experiments, conducted with a number of different cell types, serve as the basis for Steinberg's "differential adhesion" hypothesis, which offers an explanation for *in vitro* cellular rearrangements on purely physical grounds, independent of any knowledge or consideration of the chemistry of the cellular constituents involved in this behavior. It should be emphasized currently under investigation

System	References	
Sponge reaggregation	MOSCONA [159]	HUMPHREYS [108, 109]
	CURTIS [60]	MACLENNAN [141]
Other invert. Reaggregation	GUIDICE [199]	GROSS [116, 117]
	FOX [7]	GIERER et al. [85]
Reaggregation and cell sorting	HOLTFRETER [224]	MOSCONA [79]
	STEINBERG [216]	ROTH and WESTON [187]
Growth control (in vitro)	EAGLE [445, 45]	RUBIN [189, 209]
	STOKER [219, 210]	DULBECCO [66, 67]
	BURGER [33]	HOLLEY [97, 177]
	SACHS [191]	
Nerve-muscle interaction (in vitro)	CRAIN [51]	SHIMADA [208]
	FISCHBACH [70]	

that this theory does not exclude the possibility that specific chemical interactions may be important in cellular interactions.

LILIEN [135–138] and MOSCONA and GARBER [78, 80] have demonstrated that chick neural retina and mouse cerebrum cells release components into the growth medium which specifically enhance the rate and/or amount of reaggregation of dissociated cells from the homologous tissue. A component with similar activity is released during the dissociation of sponge cells, which also selectively enhances aggregation of homologous cells [105, 159, 225]. Some aggregation-promoting activity is also obtained from several types of chick embryo and mammalian cell cultures, but, in this case, no specificity is observed and the primary active component appears to be hyaluronic acid [178]. Although work on these factors is still in its initial stages, it is hoped that further studies in this area will help to elucidate the chemical basis of the cell interactions involved in tissue organization.

The *in vitro* reaggregation procedures pioneered by MOSCONA have also been employed in studies of mouse neural tissue by DE LONG and SIDMAN. DE LONG [62] demonstrated that long-term aggregates of cells from several different regions of the brain have a close histological similarity to the same regions in the intact brain. DE LONG and SIDMAN [63] then examined the histology of aggregates obtained from normal and *reeler* mutant brains. This mutation is a neurogenic mutant in which the final position of the neurons is abnormal in many regions of the brain [46, 211, 212]. When the aggregates obtained from *reeler* cells *in vitro* were subjected to histological analysis, it was found that the organization of these aggregates was very similar to the abnormal neuronal distribution in the intact mutant brain [63]. In a separate study, electron microscopic examination revealed myelination and morphological structures resembling synapses in reaggregates of cells from normal brains [200]. These studies extend MOSCONA's initial and STEINBERG's subsequent finding that the final organization of the aggregate indicates cellular recognition at the organ level to another level: recognition of different cell types within the same organ.

Fox and his co-workers have adapted reaggregation studies to *Drosophila* [72, 99, 100, 140]. Cells obtained from early embryos have been shown to form aggregates which, when kept in stationary culture, undergo organization and apparent cellular differentiation [7, 140]. Antiserum against whole cells blocks aggregation and precipitates a protein removed from the cell surface by hypotonic shock [7]. Although there is no direct demonstration that this protein is a reaggregation "factor" analogous to those described above, it is certainly involved in the reaggregation process. The ease of application of this system to genetic analysis should make it a powerful tool in the study of cell interaction.

An interpretation of the above studies should be made only in full awareness that at least three cellular processes are being observed: recovery of the cells from the dissociation procedure, formation of initial contacts and eventual establishment of final cell contacts. Each of these processes has served as the foundation for a separate theoretical explanation for cell sorting during reaggregation. Curti's timing hypothesis [57, 58] attempts to explain the species-specific reaggregation of sponge cells (discussed in detail below) as the result of the reappearance of cohesive molecules at different times after dissociation in different species of sponges. In this case, at any given time of observation, there would be two populations of cells with respect to surface adhesiveness. The more adhesive population could be expected to form tighter, more compact reaggregates and to serve as collection sites for the second cell type [59]. Steinberg [216] and Roth and Weston [187] also explain reaggregation as the result of populations of cells which differ in their adhesiveness. However, the emphasis of these workers differs significantly. Steinberg examines the relative position of two cell types in cell clusters resulting from long-term reaggregation of mixed cell populations and from apposition of preformed aggregates of different cell types [216, 246]. From such studies he has concluded that the adhesiveness of cells from different tissues can be placed in a hierarchy (cell A is more adhesive than B and both are more adhesive than C, etc.) [216, 246] and has successfully tested the validity of this hierarchy by a purely physical demonstration of the strengths of adhesion between cells of the same type [179].

These procedures demonstrate the final adhesiveness of the two cell types and thus provide information on the relative strength of cellular adhesions. However, these results concerning the relative adhesiveness of some types of cells differ in some aspects from those of Roth and Weston [187] and Roth [186]. These investigators study reaggregation by observing the rate and extent to which two preformed aggregates of different cell types adsorb cells from a single-cell suspension containing cells of only one type. Such studies determine the importance of the initial intercellular contacts formed, since the technique depends on the probability that effective cellular adhesion will occur. Only in those cases where initial and final cell contacts have the same relative strength should the results of ROTH and WESTON and STEINBERG be expected to be in agreement.

The three different processes involved in reaggregation are, as mentioned above, recovery of the cells from the dissociation procedure, formation of initial contacts, and establishment of final cell contacts. Recent reports indicate that the result of any sorting-out experiment is dependent on the cumulative effects of all three of these processes [8, 246]. In these studies, it was found that the final position of cells in aggregates of two cell types can be reversed by altering the dissociation procedure, and that the adhesive properties of at least one cell type changed during reaggregation. While sorting-out experiments have yet to provide a biochemical explanation for cell interaction, the results of these studies will have to be considered in any detailed explanation of the biochemical basis of cell interaction.

Sponge Reaggregation

Reaggregation of dissociated sponge cells serves as a useful model system for the analysis of cell interactions. WILSON first reported that species-specific reconstitution of a complete sponge could be obtained by permitting a suspension of single cells to reaggregate [244, 245]. This observation was extended by GALTSOFF [74, 75], who demonstrated that divalent cations are necessary for reaggregation [76]. HUMPHREYS used GALTSOFF's observation to obtain a soluble factor from the single-cell suspension which enhanced the ability of the cells to reaggregate [112], and obtained evidence suggesting that the soluble factors were different and specific for two sponge species [104]. These observations are the foundation for subsequent attempts to determine the molecular mechanisms of cell interaction and recognition by studying sponge cell reaggregation.

This is a useful model system for at least three reasons: 1. Dissociation of the sponge into a suspension of single cells can be accomplished by very gentle means. 2. Very mild treatment (dissociation of the sponge in the absence of divalent cations) releases a factor from the surface of the sponge cell which greatly enhances the ability of the cells to reaggregate. The mild procedure used to dissociate the cells and release factor minimizes the necessity of providing explicit or implicit qualifications concerning the effect of tissue damage on the experimental results. 3. The reaggregation of sponge cells, whether mediated by factor or not, displays specificity. Therefore it is possible to use this system to determine the chemical basis of cell association

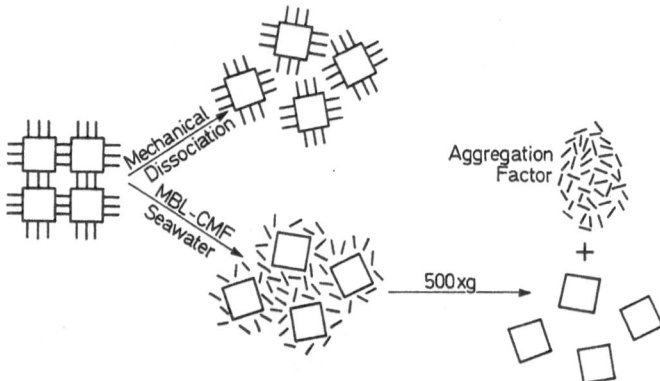

Fig. 1. Schematic representation of the results of mechanical (MD) and chemical (CMF-SW) dissociation of sponge cells. ⟱ Sponge cell with factor still attached. ☐ Sponge cell stripped of aggregation factor. ⟱ Aggregation factor

and recognition. This section of the review will describe developments in this area and present a working model for cell association which is the basis of our present studies.

In an extensive series of studies, Humphreys [105–110] has developed methods (outlined in Fig. 1) for obtaining isolated sponge cells and surface material which promotes their aggregation. Single-cell suspensions are obtained by mincing a sponge and forcing the cells through bolting silk (mechanically dissociated cells). When placed in seawater in a rotating flask, the cells will reaggregate at 22° C and at 4° C, although much less reaggregation occurs at the lower temperature. Single cells can also be obtained by washing the sponge pieces in calcium- and magnesium-free seawater. Cells prepared in this manner aggregate very slowly at 22° C and even more slowly at 4° C when the reaggregation is conducted in complete seawater. When cells prepared in calcium- and magnesium-free seawater are reaggregated in calcium- and magnesium-free seawater at either 22° C or 4° C, the extent of reaggregation is markedly reduced compared to the same cells reaggregated in seawater containing calcium. Dissociation of cells by calcium- and magnesium-free seawater results in the release of an aggregation factor (AF) into the medium. Factor activity is expressed as enhanced aggregation of calcium- and magnesium-free cells in seawater at both 4° C and 22° C. There is a requirement for Ca^{+2} and Mg^{+2}, since factor-mediated reaggregation does not occur in calcium- and magnesium-free seawater.

During these studies, Humphreys observed that two sponge species, *Microciona prolifera* and *Haliclona occulata*, displayed species specificity during reaggregation; that is, mixtures of cells of the two species reaggregated so that each aggregate contained cells of only one species, and AF from one species only enhanced reaggregation of the homologous species [105, 108–110]. These results confirmed earlier observations of species-specific reaggregation [74, 75, 245].

The basis and general applicability of *species*-specific aggregation has been questioned by CURTIS [57–59] and MacLENNON and DODD [141, 142]. These authors have reported that specific sponge reaggregation can be observed only with selected spong especies. On the basis of their immunological studies of sponge-surface components [141, 142], MacLENNON and DODD suggest that specificity of reaggregation is due to lack of any close taxonomic relationship between the species being studied. CURTIS [57, 58] originally suggested (as mentioned above) that species specificity will be observed only when the two species being investigated repair surface damage due to the dissociation procedure at different rates. More recently, however, CURTIS [59] has obtained evidence for aggregation factors which inhibit reaggregation of heterologous cells and enhance aggregation of homologous cells. This last observation was reported for cells of the same species, but different types. Recently STEINBERG and LEITH [217] observed that *Microciona* cell suspensions can be fractionated into several classes of cells which respond differently to *Microciona* aggregation factor. These observations and the chemical basis of species-specific sponge reaggregation will be reevaluated later in this review.

Crude aggregation-factor preparations consist of the $5\,000 \times g$ supernatant obtained from calcium- and magnesium-free seawater-dissociated cells [105, 107]. The factor activity is not dialyzable and is destroyed by heating to 56° C for short periods. Studies of the chemical nature of aggregation factors from a number of different species indicate that the factor is a glycoprotein [80, 143, 225]. The sugar-to-protein ratio varies from one species to another [143]. *Microciona* factor, for example, is composed of protein:hexose:uronic acid in a ratio of 49:45:6 [225]. HUMPHREYS has estimated the molecular weight by sucrose gradient centrifugation to be $5–10 \times 10^6$ [107]. Electron microscopic examination of factor preparations has provided two different results: MARGOLIASH et al. [143] has visualized 20 Å particles arranged in circular clusters, while HUMPHREYS observed a "sun-burst" pattern of fibers 45 Å in diameter [104]. The biological activity of *Microciona* factor preparations is increased by pelleting at $105\,000 \times g$ and resuspending the pellet in seawater [107, 225]. The resulting supernatant contains no detectable activity.

GALTSOFF [76] was the first to detect the lack of reaggregation in the absence of Ca^{+2} and HUMPHREYS [105] pointed out that factor preparations lose their activity if stored without Ca^{+2}. Mild EDTA treatment of AF results in loss of biological activity and more extensive treatment causes dissociation of factor which is detectable in the electron microscope [104]. These observations suggest that the Ca^{+2}-dependence of reaggregation could be due to the ability of Ca^{+2} to stabilize AF as well as the possibility that it is an essential component of the AF-cell complex. AF is degraded by some proteases [81, 225] but is resistant to the action of ribonuclease, deoxyribonuclease, hyaluronidase and lysozyme [159]. Periodate oxidation also results in destruction of AF activity [159, 225], suggesting that the carbohydrate moiety is functional, although other interpretations cannot yet be excluded. Recent investigations into the importance of the carbohydrate portion of AF have demonstrated that glucuronic acid was the most effective of the monosaccharides tested in inhibiting *Microciona* AF action in a hapten-type immunochemical inhibition assay [20, 59]. Inhibition by glucuronic acid was not observed with *Cliona* AF, implicating the carbohydrate of AF as a possible determinate of species specificity. Further evidence for the involvement of AF glucuronic acid residues in *Microciona* aggregation was obtained by the observation that glucuronidase treatment causes a loss of AF activity [37, 225].

Several attempts have been made to employ immunological methods to study species specificity of AF preparations [131, 142, 215]. MacLennan prepared cross-reacting antisera to sponge species, which also failed to show species-specific reaggregation [142]. Kuhns and Burger confirmed this observation when they produced noncrossreacting antisera against *Haliclona* and *Microciona* AF, two species which do show species-specific aggregation [131]. Spiegel [215] and Kuhns and Burger [131] have been able to demonstrate that species-specific antisera block only the reaggregation of the homologous species. However, the implicit assumption that antigenicity and AF activity are due to the same components of the preparations is not justified, since frozen and thawed sponge extracts have antigens common to factor preparations but no factor activity [141]. Finally, the antisera have all been raised against 5 000 × g supernatants, as factor preparations and enrichment of *Microciona* factor activity by pelleting at 105 000 × g is accompanied by the loss of all antigenic reactivity to the antisera against the 5 000 × g preparation [131].

Throughout this review, the term species specificity has been used to convey noncross-reaction of cells or aggregation factors from sponges of different species, families, etc. The study of sponge-cell reaggregation has been undertaken with the expectation that some basic mechanisms involved in cell interactions can be examined in detail with this system and this knowledge applied to more complex cell interactions. Clarification of the extent to which strict *species* specificity is demonstrated by sponge aggregation is required in order to determine the extent to which this expectation can be fulfilled. McClay [147] has recently studied this problem by employing the method of Roth and Weston [187] to examine aggregates formed by cell suspensions composed of two species, where one cell type has been radioactively labeled prior to mixing with preformed unlabeled aggregates of the second cell type. Autoradiographic examination of such aggregates indicates that mixtures of homotypic cells form aggregates in which the labeled and unlabeled cells

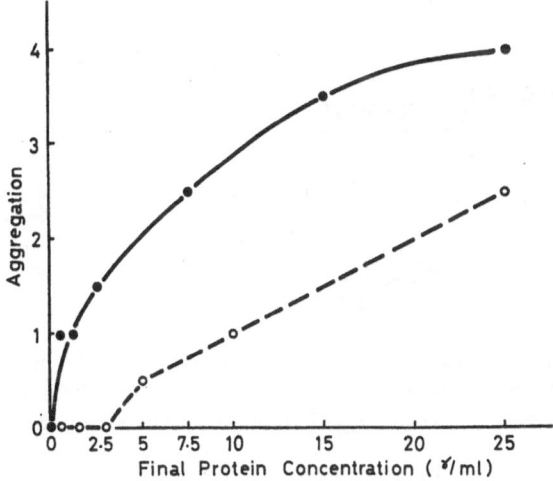

Fig. 2. Aggregation of *Microciona prolifera* cells by *Microciona prolifera* aggregation factor (——•——) and by *Haliclona occulata* aggregation factor(– –o– –). Reaggregation was assayed at 22° C in a total volume of 3 ml consisting of 1.5 ml of *Microciona* cells and 1.5 ml of aggregation factor diluted with sea water to yield the final protein concentration indicated. The original protein concentration of the *Microciona* factor preparation was 750 μg/ml and the *Haliclona* factor contained 150 μg/ml of protein, which means that adding 0.10 ml of each factor brings the final protein concentration to 5 μg/ml in the case of *Haliclona* factor and 25 μg/ml in the case of *Microciona* factor. The degree of aggregation indicated was attained in 20 minutes (at 22° C) and there was no significant increase in aggregation during the subsequent 60 minutes. 50 μg/ml of *Haliclona* factor did cause complete aggregation [4] in 20 minutes. (From TURNER et al. [225])

are intermingled, while heterotypic mixtures either result in aggregates with a sharply defined boundary or in aggregates which contain no labeled cells. This is interpreted as an indication that at least the initial cell-cell contacts are species-specific for all combinations of the five species examined. This study, in common with many others, was undertaken to determine the existence of species specificity. Such studies usually examine either the ability of a mixture of cells from two species to form separate, species specific aggregates [74, 75, 108–110, 112, 245] or the ability of crude AF (5 000 × g supernatant) from one cell type to enhance aggregation of the homologous cell type preferentially [105, 108, 109].

TURNER, WEINBAUM, and BURGER [225] have examined the ability of homologous and heterologous enriched AF preparations (105 000 × g pellets) to enhance the aggregation of one cell type (Fig. 2). These experiments employed *Microciona* cells and *Microciona* and *Haliclona* AF. These AFs have repeatedly been shown to cause species-specific aggregation [105, 108–110]. However, when aggregation of a constant number of cells was tested with increasing amounts of enriched *Microciona* and *Haliclona* AF, it was found that the *Microciona* cells aggregated with larger doses of *Haliclona* AF. Although further studies of this nature are clearly necessary, it is obvious

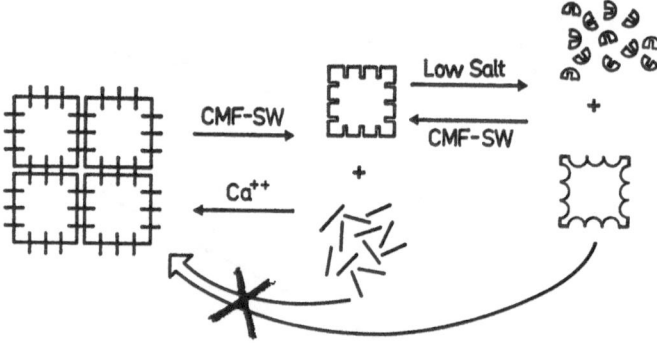

Fig. 3. Diagrammatic representation of the dissociation of the intact sponge (on the far left) into individual cells and the surface components involved in reaggregation. ⬚ Cells without factor ( ). ⬚ Factor-.less cells stripped of baseplate (ᴆᴓᴓᴓ)

that species-specific aggregation is due to a quantitative, and not a qualitative, difference in the reactivity of the cell surface with AF. Statements about the presence or absence of absolute species-specific aggregation or any other possible taxonomic relationship as the basis of specificity would appear to be premature until more is known about the nature of the interactions between the cell and the AF involved in sponge-cell aggregation.

Although high doses of AF can aggregate hetrologous cells, the existence of a quantitative difference in the ability of homologous and heterologous AF to cause aggregation indicates the presence of a system by which homologous AF is recognized. Experiments previously referred to indicate that *Microciona* cells recognize glucuronic acid covalently bound to *Microciona* AF [37, 59, 225]. WEINBAUM and BURGER [64, 236, 237] have employed hypotonic shock to release a component from the surface of *Microciona* cells previously stripped of AF in calcium- and magnesium-free seawater which has the property of a receptor for *Microciona* AF. This preparation by binding to *Microciona* AF inactivates such AF. Furthermore, cells stripped of this component by hypotonic shock can no longer be aggregated by AF unless the "shockate" is returned to the aggregation assay. This "baseplate" preparation partially inhibits the reaggregation of mechanically dissociated cells. The baseplate is sensitive to periodate, is not dialyzable and does not pellet at $105\,000 \times g$. Preliminary experiments with baseplate covalently bound to beads, as well as with factor covalently bound to beads, support the conclusion that the baseplate preparation recognized AF-bound glucuronic acid.

It is possible to draw some preliminary conclusions from the observations presented and to use these conclusions to construct a model on which to base further research. This model is clearly an *ad-hoc* model explaining our present data and will be subject to modifications as soon as new results become available. Sponge-cell aggregation is mediated by a protein-carbohydrate complex with a molecular weight of about $5-10 \times 10^6$. Both the protein and the carbohydrate portions of the factor seem to be necessary for biological activity. Ca^{+2} is necessary to maintain factor activity. Although there is much evidence for specific interaction between homologous cells and factor, in at least one species of sponge the specificity can be overcome by high concentrations of heterologous factor. The specificity of *Microciona* aggregation is probably due to recognition of AF-bound glucuronic acid by a site on the cell surface. The preliminary working model we propose is described in detail in Fig. 3.

Future research on sponge-cell aggregation will attempt to determine the chemistry of the interaction between baseplate and factor. STEINBERG and LEITH's preliminary observation [217] that different cells from one sponge reaggregate differently will undoubtedly provide a basis for detailed examinations of the extent to which sorting out between different types of cells occurs within one single species of sponge.

The knowledge gained by such studies could be applied to more complex cell interaction systems such as *in vitro* reaggregation and sorting out by vertebrate cells. We hope that future studies will be of value in elucidating the mechanism of cellular recognition involved in processes such as tissue formation during development and, perhaps, even the loss of tissue integrity which is characteristic of neoplasia.

Density-Dependent Inhibition of Growth
Introduction

The two characteristic changes in cellular activity which accompany neoplastic transformation are the loss of tissue integrity and the acquisition of the capacity for uncontrolled cell division. In more general terms, carcinogenesis is accompanied by aberrant cell-cell interaction and abnormal regulation of cellular activity, particularly growth. The initial portion of this review outlined studies of model systems concerned with understanding the process of cellular recognition, and the final sections will summarize studies of density dependent inhibition of cell division *in vitro* as a model system for understanding the regulation of cellular activity.

Transformation of tissue culture cells produces behavioral alterations with respect to cell movement and cell division. In both cases, the transformed cells behave in a manner indicating that they have lost a control mechanism which restricts the behavior of the normal cells. A nontransformed cell reacts to contact with a neighboring cell by retreating from the site of initial contact. This behavior causes a population of normal cells to have its members aligned with their long axis in parallel, giving the impression of some degree of organization. Transformed cells, on the other hand, do not retreat from the initial

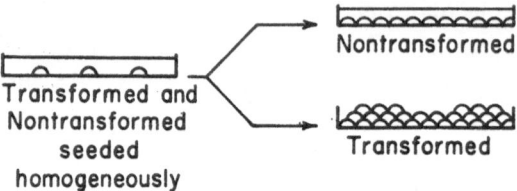

Fig. 4. Saturation density as an indicator of density-dependent inhibition of growth. Nontransformed cells stop growing after reaching a monolayer while transformed cells pile up, leading to a higher cell density

contact and show no qualms about crawling over each other, giving rise to a population of cells whose members have no predominant orientation relative to one another. This alteration in cell sociology [12] suggests that the restrictions on mobility observed in normal cells do not exist in transformed cells. This restriction was described by ABERCROMBIE and co-workers and has been termed "contact inhibition of movement" [1–3]. The second of these two behavioral alterations is that transformed cells grow to higher saturation densities than their untransformed counterparts. The nontransformed cells will divide only until the surface of the growth vessel is covered by a monolayer of cells, but the transformed cells continue to grow, forming a layer of cells which may be several cells deep (Fig. 4). The behavior of the nontransformed cells is called, in analogy to contact inhibition, density-dependent inhibition of growth [220]. The result of this changed cellular behavior is that, as in contact inhibition of cell movement, the transformed cells appear "less organized" than the nontransformed cells—apparently due in both cases to the deterioration or breakdown of control mechanisms which regulate cellular activity.

The ease with which density-dependent inhibition (DDI) of growth can be assayed makes it more amenable to investigation than contact inhibition of movement. Since the original definition by STOKER and RUBIN [220], attempts to induce DDI in transformed cells and to cause the release of DDI in nontransformed cells have succeeded in identifying several agents which can reverse the behavior characteristic of transformed or nontransformed cells. DDI of nontransformed cells has been shown to be released by increased serum concentration [97], brief treatment with very low doses of proteases [31, 201], treatment with hyaluronidase [227], colchicin [228] and increases in the pH of the medium [9, 45, 189]. It has been demonstrated [36] that transformed cells can become subject to DDI. These experiments will be discussed in more detail below and are mentioned here to point out the advantages of this system for an analysis of the regulation of cellular activity: the activity (cell division) is easily assayed and can be experimentally induced or repressed.

There are a number of possible explanations for DDI of growth by non-transformed and not by transformed cells:

1. Transformed cells may be less sensitive to decreased nutrient levels or increased levels of growth inhibitors which result from a high cell density. This explanation requires experimental conditions which allow the accumulation of "spent" medium and predicts that DDI does not require cell contact in order to be expressed. Provision of a continuous supply of fresh medium by perfusion of the culture chamber does allow transformed and nontransformed cells to grow to a higher saturation density. However, even in these conditions, the transformed cells still grow to a higher density than the non-transformed cells, indicating that the fresh medium does not prevent expression of a greater degree of density-dependent inhibition of growth by the normal cells. DULBECCO has shown that different cells in the same culture dish are subject to, or released from DDI depending only on whether they are or are not in close contact with other cells [66]. This experiment indicates that the spent medium has no effect on DDI. Finally, the assertion that DDI does not depend on cell contact has not been confirmed, since it has not been possible to transfer DDI across a Millipore filter [198].

2. The transformed cell surface may be less adhesive to the surface of the culture dish and/or to neighboring cells [50]. This decreased adhesiveness could result in greater cellular mobility and membrane flexibility, both of which could permit cytokinesis to occur more easily than in a less flexible, less mobile cell. At present, descriptive observations of cellular morphology [65] and indirect chemical evidence such as the increased turnover of membrane components [232] suggest that the transformed cell membrane is altered in a manner which may result in decreased adhesivity.

3. Contact with adjacent cells may elicit a signal which causes the onset of inhibition. Since transformed cells can be inhibited by contact with normal cells [16, 218], it seems probable that transformation results in the loss of the ability to release or transmit such a signal, but not a decreased ability to receive or respond to it.

The available evidence indicates that cell-cell contact is a prerequisite for density dependent inhibition of growth, thereby eliminating the first explanation. Elimination of this explanation does not negate the existence or importance of growth factors, but places limits on the manner in which they act; i.e., they must be assumed to be part of, or mimic part of a cellular response to cell contact which results in cessation of growth. Whether the action of such agents is only to alter the chemistry of the cell surface or whether it is to trigger a signal to stop division is as yet undetermined. Resolution of this question will allow a choice to be made between the last two alternatives. Until such a choice can be based on experimental evidence, it can only be concluded that cell contact is required for density-dependent

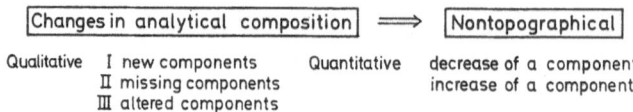

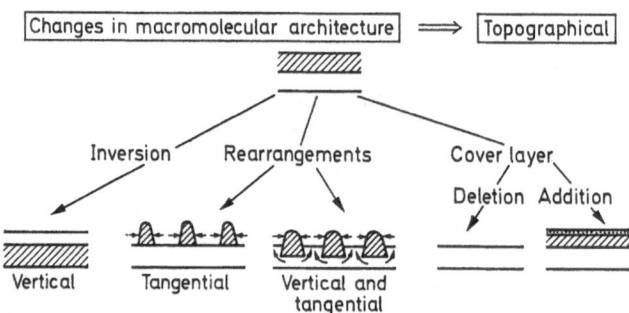

Fig. 5. Possible changes in the surface of tumor cells. Obvious overlaps between the two principal groups can occur. Changes in analytical composition can coincide with changes in architecture either with or without an obvious causal relationship existing between the two coincident events. For example, a new synthesized component may form an additional cover layer, or synthesis of new components over the entire cell-surface may occur at the same time as localized tangential rearrangements.
(From Burger [30])

inhibition of growth to occur. This requirement for cell contact indicates that growth control is mediated by the cell surface. The remainder of this review will outline studies on the cell surface which investigate its possible role in growth control.

Surface Chemistry

Fig. 5 outlines possible cell-surface alterations and suggests that these can be divided into changes in composition and structural rearrangements, which, though investigated separately, are usually interrelated and will lead to altered membrane function. We will briefly discuss growth related alterations in membrane chemistry as a prelude to a detailed description of our work on membrane structure in growth control.

The properties of transformed cells relative to the parent line have been studied as a means of understanding both transformation and growth control. Alterations in membrane carbohydrates and complex carbohydrates have been investigated in detail, while simple lipids and proteins have not been subjected to extensive analysis so far.

Ohta et al. first observed that the content of surface neuraminic acid was lower in transformed cells [171]. Wu et al. [251] extend this observation to membrane fractions of SV40-transformed 3T3 cells which contain not only much less neuraminic acid and N-acetyl-galactosamine but also reduced amounts of N-acetyl-glucosamine, fucose, mannose and galactose. Similar results were obtained for fucose, mannose and galactose in hamster cells after transformation by either polyoma virus or SV40 [27]. The most recent studies [88, 129] on neuraminic acid levels indicate, in contrast to an earlier study [130], that the level of this carbohydrate does decrease following transformation of a number of cell lines. Particularly striking is the finding [55] that SV40-3T3 revertants or variants have returned to density-dependent inhibition of growth and also have the high sialic acid content typical of the untransformed cell. It has been suggested that these decreased carbohydrate levels following transformation affect cell contacts [238], protein secretion [86], cell adhesiveness [229] and electrophoretic mobility [231], but none of these proposals have been directly supported and related to growth control. In view of the studies cited below, it is likely that the depressed levels of simple sugars reflects changes in glycolipids and glycoproteins following transformation.

RAPPORT et al. [182] and TAL [222] have reported that lactosylceramide functions as a tumor-specific hapten for antisera against tumor cells. TAL was able to obtain lactosylceramide inhibited antiserum from tumor-bearing patients and from pregnant women, which suggests that this glycolipid functions as an antigen in two types of rapidly dividing cells, tumor cells and embryonic cells. Studies of a second glycolipid antigen, the Forssman antigen, also indicate that expression of the antigen is increased in tumor and fetal cells [32, 71, 94, 169, 172]. It should be noted that BURGER [32] and HAKOMORI and KIJIMOTO [94] were able to increase Forssman reactivity of nontransformed adult cells by treating the cells with chymotrypsin [32] or EDTA [94], indicating that the antigen is present, but not reactive on the surface of normal adult cells.

HAKOMORI and MURAKAMI [93] have detected a fall in the level of hematoside after polyoma-virus transformation of BHK cells. MORA et al. [152] found that transformed 3T3 cells had a similar amount of hematoside, but less mono- and disalylgangliosides than the nontransformed cells. The apparent contradiction between the hematoside levels observed in these two studies may arise from the difference in growth conditions of the cells at the time of assay. However, both results agree in general that some glycolipid levels decrease and support the observation that human adenocarcinoma cells were altered with respect to blood-type specificity in a manner suggesting the loss of terminal sugars from glycolipid blood group determinants [91, 92, 127]. HAKOMORI and MURAKAMI [93] have suggested that the presence of completed carbohydrate chains in membrane glycolipids is necessary for density-dependent inhibition of growth. More recent studies [90, 184, 193, 233] which have examined the glycolipids of growing and non-growing nontransformed cells indicate that the growing normal cells share the incomplete glycolipids of transformed cells and thus support the proposal of HAKOMORI and MURA-KAMI [93].

Biochemical studies of glycoproteins in normal and transformed cells have provided contradictory evidence for the role of these macromolecules in growth control. MEEZAN et al. [147] found increased glucosamine incorporation into normal cells, but no attempt was made to demonstrate the nature of this material and it may be due to the greater glycolipid and simple sugar synthesis referred to above. SAKIYAMA and BURGE [192] repeated this work and found that resolution of mucopolysaccharide, glycolipid and glycoprotein by disc-gel electrophoresis indicated identical glycoprotein labeling in the normal and transformed cells. However, ONODERA and SHEININ [173] have observed that trypsin releases two glucosamine-labeled surface components from the surface of 3T3 cells which are not released from SV40-3T3 cells. This study suggests that inhibition of growth is correlated with the synthesis of additional surface components, although increased degradation of high molecular weight compounds cannot be excluded.

WARREN and his collaborators [24, 26, 233, 234] have reported that a fucose-labeled glycoprotein is present in increased amounts in DNA- or RNA-virus transformed cells of mouse, chick or hamster [24–26] and rapidly growing normal hamster cells [26] relative to the nontransformed non-growing cells. This glycopeptide is present in increased amounts in cells infected by a temperature-sensitive mutant of RSV when the cells are grown at the permissive temperature compared to the same cells grown at the restrictive temperature [233]. Their most recent results [234] report that the glycopeptide contains an additional component of sialic acid and that a sialyl transferase specific to the desialylated form of the glycopeptide is present in greater amounts in the temperature-sensitive transformed cells at the permissive than at the restrictive temperature. This same transferase also shows increased activity in transformed cells compared to the non-growing nontransformed cells. This series of experiments suggests that growing cells have a higher level of a fucose-labeled glycoprotein which is rich in sialic acid and also has a higher activity of a sialyl transferase than non-growing cells.

Glycosyl transferases, as suggested by WARREN et al. [234], may cause some or all of the changes in carbohydrate composition of the cell surface outlined above. Since most glycolipids and at least one class of glycoproteins are incomplete except in non-growing cells, explanations of these results whith assume that the incomplete carbohydrate chains reflect a synthetic defect require that the relevant transferases have a higher activity in nondividing than in dividing cells. However, the observations of WARREN suggest that the reverse is true for at least one transferase. This transferase seems to form a glycoprotein whose presence is correlated with a high growth rate. Other studies of glycoprotein transferases, using both endogenous and added acceptors, indicate that several transferases are higher in transformed cells [20, 22, 23] and growing cells [20], but that at least three other sialyl transferases have reduced activity on the transformed cell surface [56, 64, 88]. Finally, the evidence of ROTH and WHITE [188] suggests that transformed cells possess transferase activity which can act on the surface of the same cell, while transferases of nontransformed cells show a preference for the surface of adjacent cells. This last piece of evidence is offered to support the theory of ROSEMAN [185] that surface transferases are actively involved in establishing normal cell-cell contacts.

The rather confused state of the increases as well as the evidence on decreases in surface transferase activities does not offer conclusive support for ROSEMAN's theory, nor is it useful in the interpretation of alterations in the composition of complex carbohydrates in surface membranes. Further studies with purified transferases of established acceptor specificity are needed, as well as thorough examinations of the enzymes involved in degradation of membrane components. Initial observations indicated that glycosidase and protease activity is higher in non-growing than in growing cells [19, 21] and is in keeping with the relationship observed between membrane turnover and growth control.

Studies of the turnover of membrane components in growing as against non-growing cells have been conducted by two different techniques: the kinetics of isotope labeling [103, 232, 235] and changes in agglutinibility following inhibition of protein synthesis [12, 17, 111]. Both types of studies indicate that turnover is more rapid in non-growing cells. It should be noted that this result suggests that metabolic or compositional differences between growing transformed and growing nontransformed cells are not necessarily identical to differences between growing and non-growing cells, since turnover in transformed cells is more rapid than in nontransformed cells [22, 232], which at first is the opposite of the result expected on the basis of the demonstration that turnover is more rapid in stationary cells.

The studies of membrane chemistry relevant to growth control, while in need of expansion, support a few general summary statements. The relation of membrane turnover and synthetic and degradative enzyme activities to inhibition of cell division is not yet clear. The general glycoprotein content of the membrane has been reported to both rise and fall following growth inhibition, but this discrepancy could be due to the fact that the two precursors used in these studies detect different glycoproteins which do change in opposite directions. It is interesting to note that when glucosamine is used to label membrane proteins, both WARREN [232] and ONODERA and SHEININ [173] detect a transient rise in incorporation during early G-1. The studies of monosaccharide content and glycolipid complexity fit together quite well, since, with a few exceptions, both types of studies suggest that growing and transformed cells possess glycolipis with incomplete carbohydrate chains.

Surface Architecture

In order to complement this information on membrane chemistry relative to growth control, we have studied membrane-structural changes during growth control as detected by the reactions of plant lectins with the cell surface. Our interest in this system arose from the observation of Aub that wheat germ lipase seemed to specifically agglutinate tumor cells [11]. We subsequently purified an agglutinin from the lipase which reacts with polyoma-, SV40-,

Table 2. A summary of lectins tested on transformed versus nontransformed cells

Lectin	Transformed/untransformed	Purity
Wheat-germ agglutinin	+/− with exceptions [11, 35, 30 ,77, 226]	Pure [35, 162]
Concanavalin A	+/− with exceptions [119]	Pure [4]
Soybean	+/− with exceptions [205]	Pure [140]
Ricinus communis	+/− [148, 167]	Pure [138]
Lens culinaris	−/+ with exceptions [17]	Pure [100]
Phytohemagglutinin	+/− [149]	Impure [163]
Pokeweed	= [148]	Impure [163]
Dolichos biflorus	= [148]	Partially pure [69]
Ulex europaeus	= [148]	Impure [145]
Lotus tetragonolobus	= [121]	Partially pure [125]

and adenovirus-transformed cells [30, 35]. This reaction is not limited to fibroblasts transformed by DNA viruses, since Rous sarcoma virus (an RNA virus) infected fibroblasts [39] and epithelial cells trasnformed by nutritional stress [15] are also agglutinated. These studies have been expanded to include a number of different agglutinins and cell types, with the result that agglutination by plant lectins seems to be a property common to transformed cells, although there are some exceptions (see Table 2 for a summary).

The reaction of agglutinins with transformed cells is not absolutely specific, since a sufficiently high concentration of most-tumor specific agglutinins will also agglutinate the nontransformed parent cell line. This, together with other observations, suggested that the agglutinin receptor sites are present on normal cells and prompted attempts to increase the reactivity of the nontransformed cells. Degradative enzymes were screened and it was found [29] that proteases, but not lipases, glycosidases or mucopolysaccharidases, cause the same degree of agglutinability with wheat-germ agglutinin in nontransformed cells as was found for untreated transformed cells. This observation has been repaeted [119, 176] and extended to other agglutinins [119, 230].

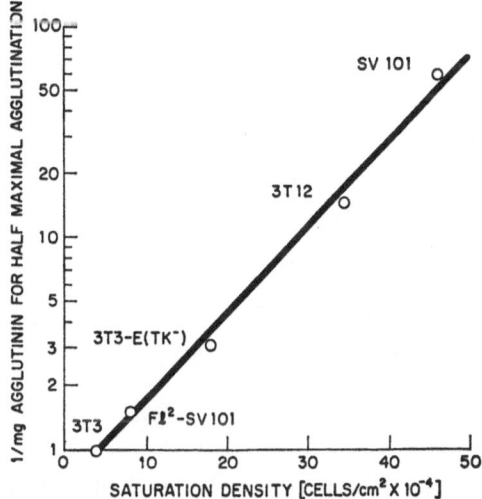

Fig. 6. Correlation between loss of density-dependent inhibition of growth and agglutinability. Adapted from data first reported by Pollack and Burger [180]

Such observations imply, but do not prove, that agglutinability is related to the growth capability of the cell. Direct support for this correlation is indicated in Fig. 6, which shows that various cell lines derived from a single stem line display increases in agglutinability which are in direct proportion to the degree to which the cells have lost density-dependent inhibition of growth [180]. This correlation is strengthened by observations that transformed cells selected for DDI lose their agglutinability [118, 180] and that cells selected for lack of agglutinability regain growth control [55, 176]. Investigations of cells infected by temperature-sensitive mutants of DNA [13, 68] or RNA [39] viruses, as well as temperature-sensitive host mutants [183], provide further evidence for a correlation between density-dependent inhibition of growth and agglutinability. In all cases, loss of growth control and increased agglutinability was expressed at the permissive temperature, and a shift to the non-permissive temperature resulted in increased growth control and decreased agglutinability.

The results outlined so far indicate that density-dependent inhibition of growth is accompanied by reduced availability of agglutinin receptor sites and that proteolytic enzymes can increase the availability of the sites. The logical extension of these observations is that proteases should release non-transformed cells from growth inhibition. This result was reported for mouse [31] and chick embryo [201] fibroblasts (Fig. 7). Control experiments indicated that the proteases did not inactivate serum components, that the effect was due to the proteolytic activity of the enzymes, and that the site of action of the proteases was the cell surface. Confluent cultures can be released from inhibition of growth several times by simply repeating the protease treatment.

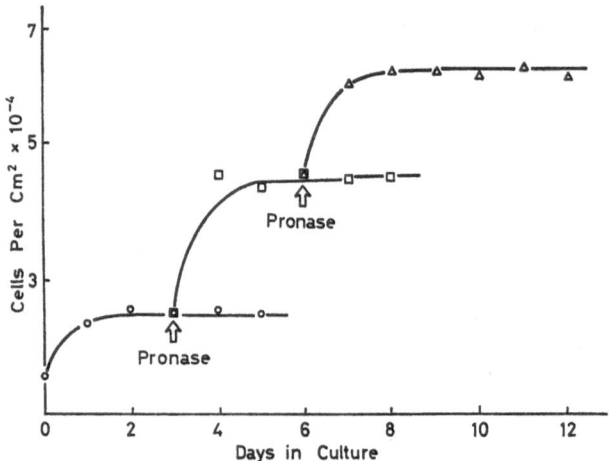

Fig. 7. Release of 3T3 cells from density-dependent inhibition of growth by treatment with trypsin. At the times indicated by the arrows, the confluent cultures were treated with 10 μg/ml of trypsin for 5 minutes. —o— untreated cells, —□— cells treated on the third day of culture, —▲— cells treated on the sixth day of culture.
(From Noonan [168])

After each treatment, the cells respond by undergoing one round of mitosis and returning to the inhibited condition. One interpretation of this observation is that the protease treatment causes a transient increase in the exposure of the agglutinin site which is involved in signaling the cells to move from G-0 into G-1 and thus initiate a round of cell division.

Studies of the agglutinability and reactivity of normal cells to fluorescent agglutinins have provided evidence which supports this interpretation [73]. Subculturing 3T3 mouse fibroblasts with trypsin provides a population of cells which is synchronous for one cell cycle. When the cells are treated in this manner and tested for agglutinability with wheat-germ agglutinin (unpublished observation) or Concanavalin A (unpublished observation) or reacted with rhodamine or fluorescein-conjugated agglutinins [73] (unpublished observations), a peak of agglutinability and fluorescence is observed which coincides with the rise in mitotic index (Fig. 8). A recent report by Shoham and Sachs [210] supports the observation that mitotic nontransformed cells are more reactive to a fluorescent agglutinin than the interphase normal cells. Analysis of the labeled mitotic cells reveals that all mitotic stages, as well as cells in early G-1, bind the fluorescent agglutinins. The fact that non-mitotic normal cells rounded up by treatment with EDTA fail to react with the fluorescent agglutinin, argues against the possibility that a positive reaction is produced simply by the shape change that the cell undergoes as it enters mitosis. This data has led us to conclude that the agglutinable state of the surface of transformed cells and protease-treated normal cells also occurs during mitosis and early G-1 of the cell cycle in nontrasnformed cells.

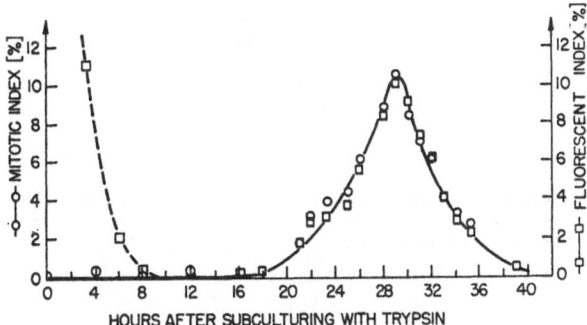

Fig. 8. Mitotic and fluorescent indices of synchronized 3T3 cells. Cells were synchronized by trypsinizing a confluent 3-day-old monolayer of 3T3 cells and replating at lower densities on cover slips. The maximum mitotic synchrony obtained (12–15%) is approximately that expected from other techniques. Cover slips were exposed to fluorescein isothiocyanate-conjugated agglutinin, fixed with ethanol, stained with Evans blue and mounted in Elvanol. In control experiments, cells were exposed to fluorescein-conjugated agglutinin and counted without fixing and staining, and fluorescence indices were identical to those reported in the figure. Blind counts of several hundred cells were made by two investigators and were quite consistent. (From Fox et al. [73•])

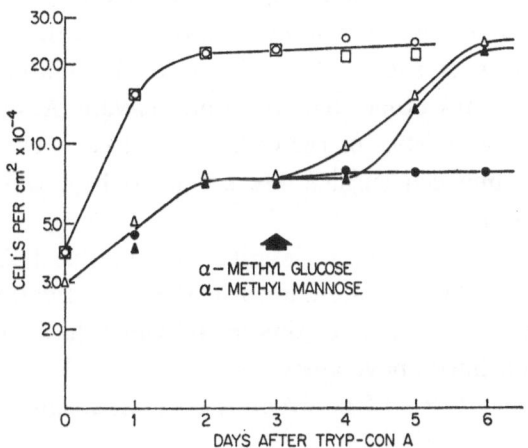

Fig. 9. Reversal of the effect of trypsinized Con A on the growth of Py 3T3 cells. —□— Py 3T3 control, —○— Py 3T3 cells + 50 µg trypsinized Con A + 0.01 M α-methyl-glucoside for 12 hours, —▼— Py 3T3 cells + 50 µg trypsinized Con A + 0.01 M α-methyl-glucoside or 0.01 M α-methyl-mannoside added to the contact-inhibited culture at day 3, —▲— Py 3T3 cells + 50 µg trypsinized Con A + 0.001 M α-methyl-glycoside or 0.001 M α-methyl-mannoside added to the contact-inhibited culture at day 3, —•— Py 3T3 cells + 50 µg trypsinized Con A. (From BURGER and NOONAN [36])

Mild protease treatment causes increased lectin agglutinability and a loss of growth control in nontransformed cells. Evidence that the agglutinable state and possibly some of the receptor sites are perhaps functional for growth regulatory processes has been provided [36]. Covering some of the agglutinin receptor sites of transformed cells by chymotrypsin-treated Concanavalin A causes the cells to become subject to inhibition of growth. This effect seemed

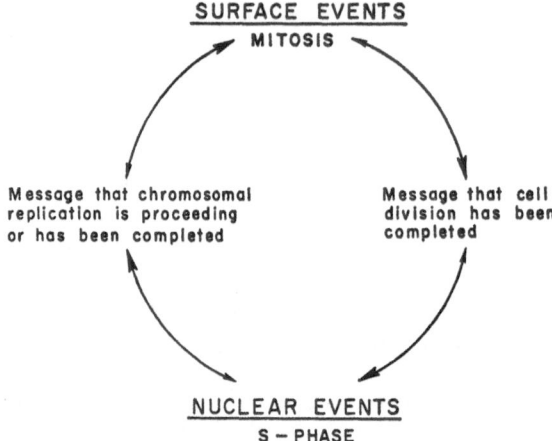

Fig. 10. Model depicting cyclic communications between the cell membrane and the cell nucleus. (From Fox et al. [73])

not to be due to cellular damage, since it could be reversed by removing the Concanavalin A from its receptor (Fig. 9). The specificity of the cellular response to the altered and presumably monovalent lectin is indicated by the fact that other proteins, such as hemoglobin, ovalbumin or two other lectins specific to surface sites other than the Concanavalin A sites, had no effect on the growth characteristics of the cells. This piece of evidence proves that an agglutinin binding cell surface site seems to be intimately involved in regulating cellular growth.

On the basis of the evidence described up to this point, we [34] have proposed a working model as a guide to further investigations of this system (Fig. 10). Essential components of this model which have been supported by the experiments outlined above are:

1. A mitotic and G-1 surface event is part of a signal that cell division has occurred.

2. The continued presence of this surface change (as in transformation or protease treatment) causes loss of growth control.

3. Masking the surface change (for example, when the sites are covered by chymotrypsin-treated Concanavalin A) results in the cessation of growth.

We have no definite information as to the nature of the signal, except that surface changes detected by lectins are at least part of it. However, we have employed this model as a basis for experiments which provide some evidence concerning its nature and the method by which it is produced.

An investigation into the possibility that endogenous surface proteases are involved in producing the uninhibited growth of tumor cells has acquired some importance. Two separate pieces of evidence support such a suggestion [29, 31]. First, addition of leukemia cells or leukemia cell membranes to nontransformed cells caused the normal cells to be released from growth

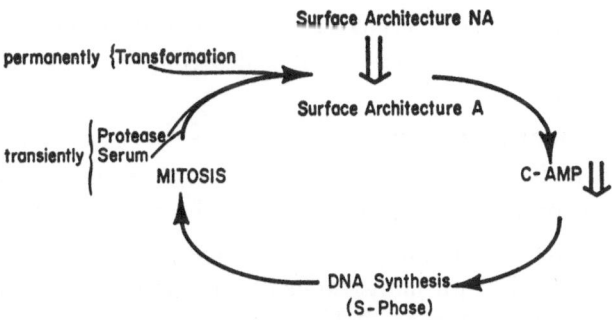

Fig. 11. Working model for the functional involvement of decreased levels of cyclic AMP in the cell cycle. NA represents the nonagglutinating form and A the agglutinating form of the cell surface architecture. The model proposes that any event causing the change in surface architecture from NA to A will also lower the cyclic AMP level at a specific point in the cell cycle, which then triggers the subsequent cell cycle. (From BURGER et al. [40])

control [33, 34]. The effectiveness of either of these treatments was abolished by the simultaneous addition of the protease inhibitor TLCK. Secondly, treatment with protease inhibitors inhibits the growth of transformed cells more effectively than that of normal cells [196, 197]. These latter studies need to be extended to prove that the site of action of the protease inhibitors is the cell surface.

Several observations suggested that cyclic AMP could be an integral element in transmitting the signal from the cell surface to the nucleus. Both growth and morphology of transformed cells reverts to that of normal cells upon the addition of dibutyryl cyclic AMP [41, 102, 124, 206]. The level of cyclic AMP in growing normal cells has been reported to be the same as [174] and higher than [207] growing transformed cells, but the confluent normal cells have a higher cyclic AMP level than confluent transformed cells [87, 174, 207]. Finally, treatment of non-growing normal cells with protease, besides causing one round of cell division and the agglutinable state, causes a drop in the intracellular cyclic AMP level [207]. These observations suggest that a decrease in the cyclic AMP pool may be a part of the signal which results in the subsequent initiation of cell division. Further support for this concept came from the following experiments. Addition of dibutyryl cyclic AMP to protease-treated cells blocked their release from density-dependent inhibition of growth [38], presumably by preventing the decrease in the intracellular cyclic AMP pool and thereby interfering with the signaling process initiated by the protease induced surface changes. It was also reported [38] that mitotic cells have a very low cyclic AMP content and that this decrease in cyclic AMP level may extend into G-1. The sequence of events implies strongly that cyclic AMP plays some part in relaying the signal from the cell surface and has led us to propose a second working model which incorporates this observation (Fig. 11). We emphasize the tentative nature of this model and realize that, like its predecessor, it will become obsolete as new observations are reported.

There are a number of unanswered questions concerning this model, whose resolution will assist us towards our goal of understanding the relationship of agglutinin receptor sites to growth control. The nature of the surface change which results in increased agglutinability is not well understood. Two possibilities, not necessarily mutually exclusive, proposed in Fig. 5, are deletion of a cover layer and tangential rearrangements. Evidence for [120] and against [9, 47, 175] the exposure of additional sites has been provided by observing the binding of radioactive lectins. A study which, unlike those reported previously, minimizes endocytosis indicates a three- to five-fold increase in binding following transformation or protease treatment [168]. "Clustering" (tangential rearrangement) of receptor sites has been detected by observing the distribution of surface labeling by ferritin-labeled agglutinins in untreated normal cells compared to transformed [144, 165] or trypsin-treated [166] cells. It is likely that both increased site exposure and clustering of the sites contribute to increased agglutinability, but the relative contributions of the two factors is undetermined. Another unclear, and so far uninvestigated, area is the mechanism which causes decreased site availability in G-1. It is possible that the increased incorporation of glucosamine into the membrane during early G-1 referred to earlier in this review [173, 232] may indicate a general synthesis of more surface components which prevent agglutinability. It is apparent that more detailed studies of membrane turnover may also provide information relevant to changes in site availability. One of the most obvious unsettled questions is the nature of the receptor site. Isolation and characterization of these sites has been initiated [5, 6, 28, 123, 164, 250], but the results are still preliminary. Resolution of this problem will allow quantitative and qualitative analysis of the differences in the sites on normal and transformed cells and thus be a significant step in solving the other open question we have mentioned, as well as providing a basis for further studies of the nature and role of cell-surface changes in the regulation of cell division.

Model systems for cell interactions involved in cell recognition and in the regulation of cellular activity are reviewed. The data presented and discussed indicate that enough is known about reaggregation of sponge cells and density-dependent inhibition of growth of tissue-culture cells to guide future research on these two systems towards the molecular level. It is also apparent that, on the basis of the information currently available, no concrete and reliable models can be presented at this stage. However, we hope the evidence and working models available will stimulate more direct and careful investigations which should lead to a better understanding of cell interactions in more complex systems as development and carcinogenesis.

References

1. ABERCROMBIE, M.: Contact inhibition in tissue culture. In Vitro **6**, 128–142 (1970).
2. ABERCROMBIE, M., HEAYSMAN, J. E. M.: Observation on the social behavior of cells in tissue culture. II. "Monolayering" of fibroblasts. Exp. Cell Res. **6**, 293–306 (1954).
3. ABERCROMBIE, M., AMBROSE, E. J.: The surface properties of cancer cells: a review. Cancer Res. **22**, 525–548 (1962).
4. AGRAWAL, B. B. S., GOLDSTEIN, I. J.: Protein-carbohydrate interaction VI isolation of concanavalin A by specific adsorption on cross-linked dextran gels. Biochim. biophys. Acta (Amst.) **147**, 262 (1970).
5. AKEDO, H., MORI, Y., TANIGAKI, Y., SHINKAI, K., MORITA, K.: Isolation of concanavalin A binding protein(s) from rat erythrocyte stroma. Biochim. biophys. Acta (Amst.) **271**, 478–487 (1972).
6. ALLEN, D., AUGER, J., CRUMPTON, M. J.: Glycoprotein receptors for concanavalin A isolated from pig lymphocyte plasma membranes by affinity chromatography in sodium deoxycholate. Nature (Lond.) New Biol. **236**, 23–25 (1972).
7. ANTLEY, R. M., FOX, A. J.: Aggregation in Drosophila. Neurosci. Res. Progr. Bull. **10** (3), 304–309 (1972).
8. AMSTRONG, P. B., NIEDERMAN, R.: Reversal of tissue position after cell sorting. Develop. Biol. **28**, 518–527 (1972).
9. ARNDT-JOVIN, P. J., BERG, P.: Quantitative binding of I^{125} concanavalin A to normal and transformed cells. J. Virol. **8**, 716–721 (1971).
10. ATTARDI, D. J., SPERRY, R. W.: Preferential selection of central pathways by regenerating optic-fibers. Exp. Neurol. **7**, 46–64 (1963).
11. AUB, J. C., TIESLAU, C., LANKESTER, A.: Reactions of normal and tumor cell surfaces to enzymes. I. Wheat-germ lipase and associated mucopolysaccharides. Proc. nat. Acad. Sci. (Wash.) **50**, 613–619 (1963).
12. BAKER, J. B., HUMPHREYS, T.: Turnover of molecules which maintain the normal surfaces of contact inhibited cells. Science **175**, 905–906 (1972).
13. BENJAMIN, T. L., BURGER, M. M.: Absence of a cell membrane alteration function in non-transforming mutants of polyoma virus. Proc. nat. Acad. Sci. (Wash.) **67**, 929–934 (1970).
14. BODIAN, D.: A model of synaptic and behavioral ontogeny, p. 129–140. In: The neurosciences: 2nd study program, editor-in-chief F. O. SCHMITT. New York: Rockefeller Univ. Press 1970. 1068 pp.
15. BOREK, C.: Neoplastic transformation *in vitro* of a clone of adult liver epithelial cells into differentiated hetatoma-like cells under conditions of nutritional stress. Proc. nat. Acad. Sci. (Wash.) **69**, 956–959 (1972).
16. BOREK, C., SACHS, L.: The difference in contact inhibition of cell replication between normal cells and cells transformed by different carcinogens. Proc. nat. Acad. Sci. (Wash.) **56**, 1705–1711 (1966).
17. BOREK, C., GROB, M., BURGER, M. M.: Surface alteration in transformed epithelial and fibroblastic cells in culture: A disturbance of membrane biosynthesis versus degradation? Exp. Cell Res. in press (1973).
18. BOON, J., TIEDEMANN, H., TIEDEMANN, H.: Inhibitors in amphibian morphogenesis: Enzymic degradation of an inhibitor for the vegetalizing factor. J. Embryol. exp. Morph. **28**, 77–86 (1972).
19. BOSMANN, H. B.: Glycoprotein degradation. Glycosidases in fibroblasts transformed by oncogenic viruses. Exp. Cell Res. **54**, 217–221 (1969).
20. BOSMANN, H. B.: Cell surface glycosyl transferases and acceptors in normal and RNA- and DNA-virus transformed fibroblasts. Biochem. biophys. Res. Commun. **48**, 523–529 (1972).
21. BOSMANN, H. B.: Elevated glycosidases and proteolytic enzymes in cells transformed by RNA tumor viruses. Biochim. biophys. Acta (Amst.) **264**, 339–343 (1972).
22. BOSMANN, H. B., HAGOPIAN, A., EYLAR, E. H.: Membrane glycoprotein biosynthesis: Changes in levels of glycosyl transferases in fibroblasts transformed by oncogenic viruses. J. Cell Physiol. **72**, 81–88 (1968).
23. BOSMANN, H. B., EYLAR, E. H.: Collagen-glucosyl transferase in fibroblasts transformed by oncogenic viruses. Nature (Lond.) **218**, 528–583 (1968).

24. Buck, C. A., Glick, M. C., Warren, L.: A comparative study of glycoproteins from the surface of control and Rous sacroma virus transformed hamster cells. Biochemistry 9, 4567–4576 (1970).
25. Buck, C. A., Glick, M. C., Warren, L.: Glycopeptides from the surface of control and virus-transformed cells. Science 172, 169–171 (1971).
26. Buck, C. A., Glick, M. C., Warren, L.: Effect of growth on the glycoproteins from the surface of control and Rous sacroma virus transformed hamster cells. Biochemistry 10, 2176–2180 (1971).
27. Buck, C. A., Glick, M. C., Hartman, J. F., Warren, L.: Presented to the 10th Int. Cancer Congr., Houston 1970.
28. Burger, M. M.: Isolation of a receptor complex for a tumour specific agglutinin from the neoplastic cell surface. Nature (Lond.) 219, 499–500 (1968).
29. Burger, M. M.: A difference in the architecture of the surface membrane of normal and virally transformed cells. Proc. nat. Acad. Sci. (Wash.) 62, 994–1001 (1969).
30. Burger, M. M.: Changes in the chemical architecture of transformed cell surfaces, p. 107–119. In: Permeability and function of biological membranes, ed. L. Bolis. Amsterdam: North-Holland Publ. 1970.
31. Burger, M. M.: Proteolytic enzymes initiating cell division and escape from contact inhibition of growth. Nature (Lond.) 227, 170–171 (1970).
32. Burger, M. M.: Forssman antigen exposed on surface membrane after viral transformation. Nature (Lond.) New Biol. 231, 125–126 (1971).
33. Burger, M. M.: The significance of surface structure changes for growth control under crowded conditions, p. 45–69. In: Ciba Foundation Symposium on Growth Control in Cell Cultures, ed. G. E. W. Wolstenholme, J. Knight. London: Churchill Livingstone 1971.
34. Burger, M. M.: Fed. Proc. (in press) (1972).
35. Burger, M. M., Goldberg, A. R.: Identification of a tumor-specific determinant on neoplastic cell surfaces. Proc. nat. Acad. Sci. (Wash.) 57, 359–366 (1967).
36. Burger, M. M., Noonan, K. D.: Restoration of normal growth by covering of agglutinin sites on tumor cell surface. Nature (Lond.) 228, 512–515 (1970).
37. Burger, M. M., Lemon, L. M., Radius, R.: Sponge aggregation. I. Are carbohydrates involved? Biol. Bull. 141, 380 (1971).
38. Burger, M. M., Bombik, B. M., Noonan, K. D.: Cell surface alterations in transformed tissue culture cells and their possible significance in growth control. J. invest. Derm. 59, 24–26 (1972).
39. Burger, M. M., Martin, G. L.: Agglutination of cells transformed by Rous sarcoma virus by wheat germ agglutinin and concanavalin A. Nature (Lond.) New Biol. 237, 9–12 (1972).
40. Burger, M. M., Bombik, B. M., Breckenridge, B. McL., Sheppard, G. R.: Growth control and cyclic alterations of cyclic AMP in the cell cycle. Nature (Lond.) New Biol. 239, 161–163 (1972).
41. Bürk, R. R.: Reduced adenylcyclase activity in a polyoma virus transformed cell line. Nature (Lond.) 219, 1272–1275 (1968).
42. Burnet, F. M.: A modification of Jerne's theory of antibody production using the concept of clonal selection. Aust. J. Sci. 20, 67 (1957).
43. Castor, L. M.: Contact inhibition of cell division and cell movement. J. invest. Derm. 59, 27–32 (1972).
44. Ceccarini, C., Eagle, H.: pH as a determinant of cellular growth and contact inhibition. Proc. nat. Acad. Sci. (Wash.) 68, 229–233 (1970).
45. Ceccarini, C., Eagle, H.: Introduction and reversal of contact inhibition of growth by pH modification. Nature (Lond.) New Biol. 233, 271 (1971).
46. Caviness, V. S., Sidman, R. L.: Olfactory structures of the forebrain in the reeler mutant mouse. J. comp. Neurol. 145, 85–104 (1972).
47. Cline, M. J., Livingston, D. C.: Binding of H^3 concanavalin A by normal and transformed cells. Nature (Lond.) New Biol. 232, 155–156 (1971).
48. Cohen, M. H.: Models for the control of development. Symp. Soc. exp. Biol. 25, 455–476 (1971).
49. Colwin, A. L., Colwin, L. H.: Role of the gamete membranes in fertilization, p. 233–279. In: Cellular membranes in development (M. Locke). New York: Academic Press 1964.

50. COMAN, D. R.: Decreased mutual adhesiveness, a property of cells from squamous cell carcinomas. Cancer Res. 4, 625–629 (1944).
51. CRAIN, S.: Bioelectric interactions between cultured fetal rodent spinal cord and skeletal muscle after innervation *in vitro*. J. exp. Zool. 173, 353–370 (1970).
52. CRICK, F. H. C.: Diffusion in embryogenesis. Nature (Lond.) 225, 420–422 (1970).
53. CRICK, F. H. C.: The scale of pattern formation. Symp. Soc. exp. Biol. 25, 429–438 (1971).
54. CULP, L. A., GRIMES, W. J., BLACK, P. H.: Contact-inhibited revertant cell lines isolated from SV40-transformed cells. I. Biologic, virologic and chemical properties. J. Cell Biol. 50, 682–690 (1971).
55. CULP, L. A., BLACK, P. H.: Contact-inhibited revertant cell lines isolated from simian virus 40-transformed cells. III. Concanavalin-A selected revertant cells. J. Virol. 9, 611–620 (1972).
56. CUMAR, F. A., BRADY, R. O., KOLODNY, E. W., McFARLAND, V. W., MORA, P. C.: Enzymatic block in the synthesis of gangliosides in DNA virus-transformed tumorigenic mouse cell lines. Proc. nat. Acad. Sci. (Wash.) 67, 757–764 (1970).
57. CURTIS, A. S.: Pattern and mechanism in the reaggregation of sponges. Nature (Lond.) 196, 245–248 (1962).
58. CURTIS, A. S. G.: Reexamination of specific sponge cell aggregation. Nature (Lond.) 226, 260–261 (1970).
59. CURTIS, A. S. G.: On the occurrence of specific adhesions between cells. J. Embryol. exp. Morph. 23, 253–272 (1970).
60. CURTIS, A. S. G., VYVER, G. VAN DE: Control of cell adhesion in a morphogenetic system. J. Embryol. exp. Morph. 26, 295–312 (1971).
61. DAN, J. C.: Morphogenetic aspects of acrosome formation and reaction. In: Advances in morphogenesis, ed. M. ABERCROMBIE, J. BRACKET, T. J. KING, vol. 8, 318 pp. New York: Academic Press 1970.
62. DeLONG, G. R.: Histogenesis of fetal mouse isocortex and hippocampus in reaggregating cell cultures. Develop. Biol. 22, 563–583 (1970).
63. DeLONG, G. R., SIDMAN, R. L.: Alignment defect of reaggregating cells in cultures of developing brains of reeler mutant mice. Develop. Biol. 22, 584–600 (1970).
64. DEN, H., SCHULTZ, A. M., BASU, M., ROSEMAN, S.: Glycosyltransferase activities in normal and polyoma-transformed BHK cells. J. biol. Chem. 246, 2721–2723 (1971).
65. DOMNINA, S. V., IVANOVA, O. G., MARGOLIS, S. B., OLSHERVSKAJA, L. V., ROVENSKY, J. A., VASILIEV, J. M., GELFAND, J. M.: Defective formation of lamellar cytoplasm by neoplastic fibroblasts. Proc. nat. Acad. Sci. (Wash.) 69, 243–252 (1972).
66. DULBECCO, R.: Behavior of tissue culture cells infected with polyoma virus. Proc. nat. Acad. Sci. (Wash.) 67, 1214–1220 (1970).
67. DULBECCO, R.: Topoinhibition and serum requirement of transformed and untransformed cells. Nature (Lond.) 227, 802–806 (1970).
68. ECKHART, W., DULBECCO, R., BURGER, M. M.: Temperature-dependent surface changes in cells infected or transformed by a thermosensitive mutant of polyoma virus. Proc. nat. Acad. Sci. (Wash.) 68, 283–286 (1971).
69. ETZLER, M. E., KABAT, E. A.: Purification and characterization of a lecitin (plant hemogglutinin) with blood group A specificity from *Dolichos biflorus*. Biochemistry 9, 869–877 (1970).
70. FISHBACH, G. D.: Synapse formation between dissociated nerve and muscle cells in low density cell cultures. Develop. Biol. 28, 407–429 (1972).
71. FOGEL, M., SACHS, L.: The induction of Forssman-antigen synthesis in hamster and mouse cells in tissue culture, as detected by the fluorescent-antibody technique. Exp. Cell Res. 34, 448–462 (1964).
72. FOX, A. S., HORIKAWA, M., LING, L.-N. L.: The use of *Drosophila* cell cultures in studies of differentiation, p. 65–84. In: In Vitro, vol. 3, Differentiation and defense mechanisms in lower organisms, ed. M. M. SIGEL. Baltimore: Williams and Wilkins 1968.
73. FOX, T. O., SHEPPARD, J. R., BURGER, M. M.: Cyclic membrane changes in animal cells: Transformed cells permanently display a surface architecture detected in normal cells only during mitosis. Proc. nat. Acad. Sci. (Wash.) 68, 244–247 (1971).
74. GALTSOFF, P. S.: The ameboid movement of dissociated sponge cells. Biol. Bull. 45, 153–161 (1923).

75. Galtsoff, P. S.: Regeneration after dissociation (an experimental study on sponges). I. Behavior of dissociated cells of *Microciona prolifera* under normal and abnormal conditions. J. exp. Zool. **42**, 183–221 (1925).

76. Galtsoff, P. S.: Regeneration after dissociation (an experimental study on sponges). II. Histogenesis of *Microciona prolifera*. J. exp. Zool. **42**, 223–251 (1925).

77. Gantt, R. R., Martin, J. I., Evans, V. J.: Agglutination of *in vitro* cultured neoplastic and non-neoplastic cell lines by a wheat germ agglutinin. J. nat. Cancer Inst. **42**, 369–373 (1969).

78. Garber, B. B., Moscona, A. A.: Enhancement of aggregation of embryonic brain cells by extracellular materials form cultures of brain cells. J. Cell Biol. **43**, (abstr.) (1969).

79. Garber, B. B., Moscona, A. A.: Reconstruction of brain tissue from cell suspensions. I. Aggregation patterns of cells dissociated from different regions of the developing brain. Develop. Biol. **27**, 217–234 (1972).

80. Garber, B. B., Moscona, A. A.: Reconstruction of brain tissue from cell suspensions. II. Specific enhancement of aggregation of embryonic cerebral cells by supernatant from homologous cell cultures. Develop. Biol. **27**, 235–243 (1972).

81. Gasic, G. J., Galanti, N. L.: Proteins and disulfide groups in the aggregation of dissociated cells of sea sponges. Science **151**, 203–235 (1966).

82. Gaze, R. M.: The formation of nerve connections. London: Academic Press 1970.

83. Gaze, R. M., Sharma, S. C.: Axial differences in the reinnervation of the optic tectum by regenerating goldfish optic nerve fibres. Exp. Brain Res. **10**, 171–181 (1970).

84. Gaze, R. M., Chung, S.-H., Keating, M. J.: The development of the retinotectal protection in *Xenopus*. Nature (Lond.) **236**, 133–135 (1972).

85. Gierer, A., Berking, S., Bode, J., David, C. N., Flick, K., Hansmann, G., Schaller, H., Trenkner, E.: Regeneration of hydra from reaggregated cells. Nature (Lond.) New Biol. **239**, 98–101 (1972).

86. Glick, J. L., Goldberg, A. R., Pardee, A. B.: The role of sialic acid in the release of proteins from L1210 leukemia cell. Cancer Res. **26**, 1774–1777 (1966).

87. Granner, D., Chase, L. R., Aurbach, G. D., Tomkins, G. M.: Tyrosine aminotransferase: enzyme induction independent of adenosine 3′,5′-monophosphate. Science **162**, 1018–1020 (1968).

88. Grimes, W. J.: Sialic acid transferases and sialic acid levels in normal and transformed cells. Biochemistry **9**, 5083–5092 (1970).

89. Grobstein, C.: Mechanism of organogenetic tissue interaction. Nat. Cancer Inst. Monogr. **26**, 279–299 (1967).

90. Hakomori, S.: Cell density-dependent changes of glycolipid concentrations in fibroblasts, and loss of this response in virus-transformed cells. Proc. nat. Acad. Sci. (Wash.) **67**, 1741–1747 (1970).

91. Hakomori, S., Jeanloz, R. W.: Isolation of a glycolipid containing fucose, galactose, glucose, and glucosamine from human cancerous tissue. J. biol. Chem. **239**, PC 3606–3706 (1964).

92. Hakomori, S., Koscielak, J., Bloch, K. J., Jeanloz, R. W.: Immunological relationship between blood group substances and a fucose-containing glycolipid of human adenocarcinoma. J. Immunol. **98**, 31–38 (1967).

93. Hakomori, S., Murakami, W. T.: Glycolipids of hamster fibroblasts and derived malignant-transformed cell lines. Proc. nat. Acad. Sci. (Wash.) **59**, 254–261 (1968).

94. Hakomori, S., Kijimoto, S.: Forssman reactivity and cell contacts in cultured hamster cells. Nature (Lond.) New Biol. **239**, 87–88 (1968).

95. Hamburger, V.: Emergence of nervous coordination. Origins of integrated behavior, p. 251–271. In: The emergence of order in developing systems, ed. M. Locke. New York: Academic Press 1968. 350 pp.

96. Hartman, J. F., Gwatkin, R. B. L., Hutchison, C. F.: Early contact interactions between mammalian gametes *in vitro*: evidence that the vitellus influences adherence between sperm and zona pellucida. Proc. nat. Acad. Sci. (Wash.) **69**, 2767–2769 (1972).

97. Holley, R. W., Kiernan, J. A.: "Contact inhibition" of cell division in 3T3 cells. Proc. nat. Acad. Sci. (Wash.) **60**, 300–304 (1968).

98. HOLLEY, R. W., KIERNAN, J. A.: Growth control in cell cultures, p. 3–15. In: Symposium on Growth Control in Cell Cultures, ed. G. E. W. WOLSTENHOLME, J. KNIGHT. London: Churchill Livingstone 1971. 275 pp.

99. HORIKAWA, M., FOX, A. S.: Culture of embryonic cells of *Drosophila melanogaster in vitro*. Science **145**, 1437–1439 (1964).

100. HORIKAWA, M., LING, L.-N., FOX, A. S.: Long-term culture of embryonic cells of *Drosophila melanogaster*. Nature (Lond.) **210**, 183–185 (1966).

101. HOWARD, I. K., SAGE, H. J.: Isolation and characterization of a phytohemagglutinin from the lentil. Biochemistry **8**, 2436–2441 (1969).

102. HSIE, A. W., PUCK, T. T.: Morphological transformation of Chinese hamster cells by dibutyryl adenosine cyclic 3′:5′-monophosphate and testosterone. Proc. nat. Acad. Sci. (Wash.) **68**, 358–361 (1971).

103. HUGHES, R. C., SANFORD, B., JEANLOZ, R. W.: Regeneration of the surface glycoproteins of a transplantable mouse tumor cell after treatment with neuraminidase. Proc. nat. Acad. Sci. (Wash.) **69**, 942–945 (1972).

104. HUMPHREYS, S.: Abstract 133, presented at the 11th Annual Meeting of Amer. Soc. of Cell Biol. (1971).

105. HUMPHREYS, T.: Chemical dissolution and *in vitro* reconstruction of sponge cell adhesions. I. Isolation and functional demonstration of the components involved. Develop. Biol. **8**, 27–47 (1963).

106. HUMPHREYS, T.: Aggregation of chemically dissociated sponge cells in the absence of protein synthesis. J. exp. Zool. **160**, 235–240 (1965).

107. HUMPHREYS, T.: Cell surface components participating in aggregation: evidence for a new cell particulate. Exp. Cell Res. **40**, 539–543 (1965).

108. HUMPHREYS, T.: The cell surface and specific cell aggregation, p. 195–210. In: The specificity of cell surfaces, ed. E. DAVIS, L. WARREN. Englewood, N.Y.: Prentice Hall 1967.

109. HUMPHREYS, T.: Biochemical analysis of sponge cell aggregation. Symp. Zool. Soc. Found. **25**, 325–334 (1969).

110. HUMPHREYS, T.: Species specific aggregation of dissociated sponge cells. Nature (Lond.) **228**, 685–686 (1970).

111. HUMPHREYS, T.: Cell contact, contact inhibition of growth and the regulation of macromolecular metabolism, p. 264–276. In: Cell interactions, 3rd Lepetit Colloquium, ed. S. G. SILVESTRI. 1971. 314 pp.

112. HUMPHREYS, T., HUMPHREYS, S., MOSCONA, A. A.: A procedure for obtaining completely dissociated sponge cells. Biol. Bull. **119**, 294 (1960).

113. HUMPHREYS, T., HUMPHREYS, S., MOSCONA, A. A.: Rotation-mediated aggregation of dissociated sponge cells. Biol. Bull. **119**, 295 (1960).

114. HUNT, R. K., JACOBSON, M.: Developmental and stability of positional information in *Xenopus* retinal ganglion cells. Proc. nat. Acad. Sci. (Wash.) **69**, 780–783 (1972).

115. HUNT, R. K., JACOBSON, M.: Specification of positional information in retinal ganglion cells of *Xenopus*: stability of the specified state. Proc. nat. Acad. Sci. (Wash.) **69**, 2860–2864 (1972).

116. HYNES, R. O., GREENHOUSE, G. A., MINKOFF, R., GROSS, P. R.: Properties of the three cell types in sixteen-cell sea urchin embryos: RNA synthesis. Develop. Biol. **27**, 457–478 (1972).

117. HYNES, R. O., RAFF, R. A., GROSS, P. R.: Properties of the three cell types in sixteen-cell sea urchin embryos aggregation and microtubule protein synthesis. Develop. Biol. **27**, 150–165 (1972).

118. INBAR, M., RABINOWITZ, Z., SACHS, L.: The formation of variants with a reversion of properties of transformed cells. III. Reversion of the structure of the cell surface membrane. Int. J. Cancer **4**, 690–696 (1969).

119. INBAR, M., SACHS, L.: Interaction of the carbohydrate-binding protein concanavalin A with normal and transformed cells. Proc. nat. Acad. Sci. (Wash.) **63**, 1418–1425 (1969).

120. INBAR, M., SACHS, L.: Structural differences in sites on the surface membrane of normal and transformed cells. Nature (Lond.) **233**, 710–712 (1969).

121. INBAR, M., VLODAVSKY, I., SACHS, L.: Availability of L-fucose-like sites on the surface membrane of normal and transformed mammalian cells. Biochim. biophys. Acta (Amst.) **255**, 703–708 (1972).

122. Jacobson, M.: Developmental neurobiology. New York: Holt, Rinehart and Winston 1970.
123. Jansons, V. K., Burger, M. M.: Isolation and characterization of agglutinin receptor sites. II. Isolation and partial purification of surface membrane receptors for wheat germ agglutinin. Biochim. biophys. Acta (Amst.) (in press) (1972).
124. Johnson, G. S., Friedman, R. M., Pastan, I.: Restoration of several morphological characteristics of normal fibroblasts in sarcoma cells treated with adenosine-3':5'-cyclic monophosphate and its derivatives. Proc. nat. Acad. Sci. (Wash.) **68**, 425–429 (1971).
125. Kalb, A. J.: The separation of three L-fucose-binding proteins of *Lotus tetragonolobus*. Biochim. biophys. Acta (Amst.) **168**, 532–536 (1968).
126. Kalckar, H. M.: Galactose metabolism and cell "sociology". Science **150**, 305–313 (1968).
127. Kay, H. E. M., Wallace, D. M.: A and B antigens of tumors arising from urinary epithelium. J. nat. Cancer Inst. **26**, 1349–1365 (1961).
128. Kocher-Becker, U., Tiedemann, H.: Induction of mesodermal and endodermal structures and primordial germ cells in *Triturus* ectoderm by a vegetalizing factor from chick embryos. Nature (Lond.) **233**, 65–66 (1971).
129. Kornfeld, S.: Decreased phytohemagglutinin receptor sites in chronic lymphocytic leukemia. Biochim. biophys. Acta (Amst.) **192**, 542–545 (1971).
130. Kraemer, P. M.: Sialic acid of mammalian cell lines. J. Cell Physiol. **67**, 23–34 (1966).
131. Kuhns, W. J., Burger, M. M.: (in preparation) 1972.
132. Lawrence, P. A.: The organization of the insect segment. Symp. Soc. exp. Biol. **25**, 379–390 (1971).
133. Lawrence, P., Crick, F. H. C., Munro, M.: J. Cell Sci. (in press) (1972).
134. Lilien, J. E.: Enhancement of the aggregation of embryonic chick neural retina cells by a supernatant prepared from monolayers of homologous cells. Ph.D. Thesis, University of Chicago 1967.
135. Lilien, J. E.: Specific enhancement of cell aggregation *in vitro*. Develop. Biol. **17**, 657–678 (1968).
136. Lilien, J. E.: Toward a molecular explanation for specific cell adhesion, p. 169–196. In: Current topics in developmental biology, ed. A. A. Moscona, A. Monroy. New York: Academic Press 1969. 233 pp.
137. Lilien, J. E., Moscona, A. A.: Cell aggregation: its enhancement by a supernatant from cultures of homologous cells. Science **157**, 70–72 (1967).
138. Lin, J. Y., Tserno, K. Y., Chen, C. C., Lin, L. T., Tung, T. C.: Abrin and ricin: new anti-tumour substances. Nature (Lond.) **227**, 292–293 (1970).
139. Ling, L.-N. L., Horikawa, M., Fox, A. S.: Aggregation of dissociated cells from *Drosophila* embryos. Develop. Biol. **22**, 264–281 (1970).
140. Lis, H., Sharon, N., Katchalski, E.: Soybean hemagglutinin, a plant glycoprotein. I. Isolation of a glycopeptide. J. biol. Chem. **241**, 684–689 (1966).
141. MacLennan, A. P.: Polysaccharides from sponges and their possible significance in cellular aggregation. Symp. Zool. Soc. London **25**, 299–324 (1969).
142. MacLennan, A. P., Dodd, R. Y.: Promoting activity of extracellular materials on sponge cell reaggregation. J. Embryol. exp. Morph. **17**, 473–480 (1967).
143. Margoliash, E., Schenck, J. R., Hargie, M. P., Burokas, S., Richter, W. R., Barlow, G. H., Moscona, A. A.: Characterization of specific cell aggregating materials from sponge cell. Biochem. biophys. Res. Commun. **20**, 383–388 (1965).
144. Martinez-Palomo, A., Wicker, R., Bernhard, W.: Ultrastructure detection of concanavalin surface receptors in normal and polyoma-transformed cells. Int. J. Cancer **9**, 676–684 (1972).
145. Matsumoto, I., Osawa, T.: Purification and characterization of a *Cytisus*-type anti-H(O) phytohemagglutinin from *Ulex europeus* seeds. Arch. Biochem. Biophys. **140**, 484–491 (1970).
146. McClay, D. R.: An autoradiographic analysis of the species specificity during sponge cell reaggregation. Biol. Bull. **141**, 319–330 (1971).
147. Meezan, E., Wu, H. C., Black, P. H., Robbins, P. W.: Comparative studies on the carbohydrate-containing membrane components of normal and virus-trans-

formed fibroblasts. Separation of glycoproteins and glycopeptides by Sephadex chromatography. Biochemistry **8**, 2518–2524 (1969).

148. METZ, C.: Senior thesis, Princeton University 1972.

149. METZ, C., MONROY, A. (eds.): Fertilization. New York: Academic Press 1969.

150. MITCHISON, N. A.: Control of the immune response by events at the lymphocyte surface. In Vitro **7**, 88–94 (1971).

151. MITCHISON, N. A.: Cell cooperation in the immune response. The hypothesis of an antigen presentation mechanism. Immunopathology **6**, 52 (1971).

152. MORA, P. T., BRADY, R. O., BRADLEY, R. M., McFARLAND, V. W.: Gangliosides in DNA virus-transformed and spontaneously transformed tumorigenic mouse cell lines. Proc. nat. Acad. Sci. (Wash.) **63**, 1290–1296 (1969).

153. MOSCONA, A. A.: Development of heterotypic combinations of dissociated embryonic chick cells. Proc. Soc. exp. Biol. (N.Y.) **92**, 410–416 (1956).

154. MOSCONA, A. A.: The development *in vitro* of chimeric aggregates of dissociated embryonic chick and mouse cells. Proc. nat. Acad. Sci. (Wash.) **43**, 184–194 (1957).

155. MOSCONA, A. A.: Patterns and mechanisms of tissue reconstruction from dissociated cells, p. 45–70. In: Developing cell systems and their control, ed. D. RUDNICK. New York: Ronald Press 1960.

156. MOSCONA, A. A.: Rotation-mediated histogenetic aggregation of dissociated cells: a quantifiable approach to cell interactions *in vitro*. Exp. Cell Res. **22**, 455–475 (1961).

157. MOSCONA, A. A.: Analysis of cell recombinations in experimental synthesis of tissues *in vitro*. J. cell. comp. Physiol., Suppl. I **60**, 65–80 (1962).

158. MOSCONA, A. A.: Cellular interactions in experimental histogenesis. Int. Rev. exp. Path. **1**, 371–428 (1962).

159. MOSCONA, A. A.: Studies on cell aggregation: Demonstration of material with selective cell-binding activity. Proc. nat. Acad. Sci. (Wash.) **49**, 742–747 (1963).

160. MOSCONA, A. A., MOSCONA, M. H.: The dissociation and aggregation of cells from organ rudiments of the early chick embryo. J. Anat. (Lond.) **86**, 287–301 (1952).

161. MOSCONA, M. H., MOSCONA, A. A.: Control of differentiation in aggregates of embryonic skin cells: suppression of feather morphogenesis by cells from other tissues. Develop. Biol. **11**, 402–423 (1965).

162. NAGATA, Y., BURGER, M. M.: Wheat germ agglutinin. Isolation and crystallization. J. biol. Chem. **247**, 2248–2250 (1972).

163. NAGATA, Y., GOLDBERG, A. R., BURGER, M. M.: The isolation and purification of wheat germ agglutinin and other agglutinins. In: Methods in enzymology, ed. S. P. COLOWICK, N. O. KAPLAN. New York: Academic Press (in press).

164. NERI, G., WALBORG, E. F., JR.: Concanavalin A and wheat germ agglutinin receptor activity of glycopeptides isolated from the surface of normal and neoplastic rat liver cells. Abstract No 35, 164th Amer. Chem. Soc. Meeting, Div. of Carbohydrate Chem. 1972.

165. NICOLSON, G. L.: Difference in topology of normal and tumour cell membranes shown by different surface distribution of ferritin-conjugated concanavalin A. Nature (Lond.) New Biol. **233**, 244–246 (1971).

166. NICOLSON, G. L.: Topography of membrane concanavalin A sites modified by proteolysis. Nature (Lond.) New Biol. **239**, 193–197 (9172).

167. NICOLSON, G. L., BLAUSTEIN, J.: The interaction of *Ricinus communis* agglutinin with normal and tumor cell surfaces. Biochim. biophys. Acta (Amst.) **266**, 543–547 (1972).

168. NOONAN, K. D.: Ph.D. thesis, Princeton Univ. 1972.

169. NOONAN, K. D., BURGER, M. M.: Architectural changes of embryonic, normal and transformed cell surfaces demonstrated by plant agglutinins. Proc. 1st Conference and Workshops on Embryonic and Fetal Antigens in Cancer, p. 59–69, eds. N. G. ANDERSON, J. H. COGGIN. 1971. 400 pp.

170. NOSSAL, G. J. V., ADA, G. F.: Antigens, lymphoid cells and the immune response. New York: Academic Press 1971. 324 pp.

171. OHTA, N., PARDEE, A. B., McAUSLON, B. R., BURGER, M. M.: Sialic acid contents and controls of normal and malignant cells. Biochim. biophys. Acta (Amst.) **158**, 98–102 (1968).

172. O'NEILL, C. H.: An association between viral transformation and Forssman antigen detected by immune adherence in cultured BHK 21 cells. J. Cell Sci. **3**, 405–422 (1968).

173. Onodera, K., Sheinin, R.: Macromolecular glucosamine-containing component of the surface of cultivated mouse cells. J. Cell Sci. **7**, 337–355 (1970).

174. Otten, J., Johnson, G. L., Paston, J.: Cyclic AMP levels in fibroblasts: relationship to growth rate and contact inhibition of growth. Biochem. biophys. Res. Commun. **44**, 1192–1198 (1971).

175. Ozanne, B., Sambrook, J.: Binding of radioactively labelled concanavalin A and wheat germ agglutinin to normal and virus-transformed cells. Nature (Lond.) New Biol. **232**, 156–160 (1971).

176. Ozanne, B., Sambrook, J.: Isolation of lines of cells resistant to agglutination by concanavalin A from 3T3 cells transformed by SV 40. In: The biology of oncogenic viruses, 2nd Int. Lepetit Colloquium, ed. E. Verwey. Amsterdam: North-Holland Publ. 1971. 339 pp.

177. Paul, D., Leffert, H., Sato, G., Holley, R. W.: Stimulation of DNA and protein synthesis in fetal-rat liver cells by serum from partially hepatectomized rats. Proc. nat. Acad. Sci. (Wash.) **69**, 374–377 (1972).

178. Ressac, B., Defendi, V.: Cell aggregation: Role of acid mucopolysaccharides. Science **175**, 898–900 (1972).

179. Phillips, H. M., Steinberg, M. S.: Equilibrium measurements of embryonic chick cell adhesiveness. I. Shape equilibrium in centrifugal fields. Proc. nat. Acad. Sci. (Wash.) **64**, 121–127 (1969).

180. Pollack, R. E., Burger, M. M.: Surface-specific characteristics of a contact inhibited cell line containing the SV40 viral genome. Proc. nat. Acad. Sci. (Wash.) **62**, 1074–1076 (1969).

181. Raff, M. C., Petris, S. de: Antigen-antigen reaction at the lymphocyte surface: implications for membrane structure, lymphocyte activation and tolerance induction, p. 237–246. In: Cell Interactions, ed. S. G. Silvestri. Amsterdam: North-Holland Publ. Co. 1972. 314 pp.

182. Rapport, M. M., Graf, L., Skipski, V. P., Alonzo, N. F.: Immunochemical studies of organ and tumor lipids. VI. Isolation and properties of cytolipin H. Cancer (Philad.) **12**, 438–445 (1959).

183. Renger, H. C., Basilico, C.: Mutation causing temperature-sensitive expression of cell transformation by a tumor virus. Proc. nat. Acad. Sci. (Wash.) **69**, 109–114 (1972).

184. Robbins, P. W., Macpherson, J. A.: Glycolipid synthesis in normal and transformed animal cells. Proc. roy. Soc. B **177**, 49–58 (1971).

185. Roseman, S.: The synthesis of complex carbohydrates by multiglycosyltransferases and their potential function in intercellular adhesion. Chem. Phys. Lipids **5**, 270–297 (1970).

186. Roth, S. A.: Studies on intercellular adhesive selectivity. Develop. Biol. **18**, 602–631 (1968).

187. Roth, S. A., Weston, J. A.: The measurement of intercellular adhesion. Proc. nat. Acad. Sci. (Wash.) **58**, 974–980 (1967).

188. Roth, S. A., White, D.: Intercellular contacts and cell-surface galactosyl transferase activity. Proc. nat. Acad. Sci. (Wash.) **69**, 485–489 (1972).

189. Rubin, H.: Growth regulation in cultures of chick embryo fibroblasts, p. 127–149. In: Growth control in cell cultures. A Ciba Foundation Symposium, ed. G. E. W. Wolstenholme, J. Knight. London: Churchill Livingston 1971.

190. Rubin, L., Saunders, J. W., Jr.: Ectodermal-mesodermal interactions in the growth of limb buds in the chick embryo: constancy and temporal limits of the ectodermal induction. Develop. Biol. **28**, 94–112 (1972).

191. Sachs, L.: The mechanism of carcinogenesis, p. 118–128. In: Molecular bioenergetics and macromolecular biochemistry, ed. H. H. Weber. Berlin-Heidelberg-New York: Springer 1972.

192. Sakiyama, H., Burge, B. W.: Comparative studies of the carbohydrate-containing components of 3T3 and simian virus 40 transformed 3T3 mouse fibroblasts. Biochemistry **11**, 1366–1377 (1972).

193. Sakiyama, H., Gross, S. K., Robbins, P. W.: Glycolipid synthesis in normal and virus transformed hamster cell lines. Proc. nat. Acad. Sci. (Wash.) **69**, 872–876 (1972).

194. Saxén, L.: Inductive interactions in kidney development. Symp. Soc. exp. Biol. **25**, 207–222 (1971).

195. SAXÉN, L., KUHONEN, J.: Inductive tissue interactions in vertebrate morphogenesis. Int. Rev. exp. Path. **8**, 57–128 (1969).
196. SCHNEBLI, H. P.: A protease-like activity associated with malignant cells. Schweiz. med. Wschr. **102**, 1194–1197 (1972).
197. SCHNEBLI, H. P., BURGER, M. M.: Selective inhibition of growth of transformed cells by protease inhibitors. Proc. nat. Acad. Sci. (Wash.) (in press) (1972).
198. SCHÜTZ, L., MORA, P. T.: The need for direct cell contact in "contact" inhibition of cell division in culture. J. cell. Physiol. **71**, 1–6 (1968).
199. SCONZO, G., PIRRONE, A. M., MUTOLO, V., GUIDICE, G.: Synthesis of ribosomal RNA in disaggregated cells of sea urchin embryos. Biochim. biophys. Acta (Amst.) **199**, 441–446 (1970).
200. SEEDS, N. W., VATTER, A. E.: Synaptogenesis in reaggregating brain cell culture. Proc. nat. Acad. Sci. (Wash.) **68**, 3219–3222 (1971).
201. SEFTON, B. M., RUBIN, H.: Release from density dependent growth inhibition by proteolytic enzymes. Nature (Lond.) **227**, 843–845 (1970).
202. SHEFFIELD, J. B.: Studies on aggregation of embryonic cells: initial cell adhesions and the formation of intercellular junctions. J. Morph. **132**, 245–264 (1970).
203. SHEFFIELD, J. B., MOSCONA, A. A.: Early stages in the reaggregation of embryonic chick neural retina cells. Exp. Cell Res. **57**, 462–466 (1969).
204. SHEININ, R., ONODERA, K.: Studies on the biochemical properties of surface components of normal and SV-40 transformed 3T3 mouse cells. Canad. J. Biochem. **48**, 851–857 (1970).
205. SELA, B., LIS, H., SHARON, N., SACHS, L.: Different locations of carbohydrate-containing sites in the surface membrane of normal and transformed cells. J. Membrane Biol. **3**, 267–279 (1970).
206. SHEPPARD, J. R.: Restoration of contact-inhibited growth to transformed cells by dibutyryl adenosine 3′:5′-cyclic monophosphate. Proc. nat. Acad. Sci. (Wash.) **68**, 1316–1320 (1971).
207. SHEPPARD, J. R.: Difference in the cyclic adenosine 3′,5′-monophosphate levels in normal and transformed cells. Nature (Lond.) New Biol. **236**, 14–16 (1972).
208. SHIMADA, Y., KANO, M.: Formation of neuromuscular junctions in embryonic cultures. Arch. Histol. Japan **33**, 95–114 (1971).
209. SHODELL, M., RUBIN, H., GERHART, J.: Nautralization of growth-inhibitory material present in calf serum by conditioning factors elaborated by chick embryo cells in culture. Exp. Cell Res. **74**, 375–382 (1972).
210. SHOHAM, J., SACHS, L.: Differences in the binding fluorescent concanavalin A to the surface membrane of normal and transformed cells. Proc. nat. Acad. Sci. (Wash.) **69**, 2479–2482 (1972).
211. SIDMAN, R. L.: Abnormal cell migrations in developing brains of mutant mice, p. 40–49. In: Expanding concepts in mental retardation. A Symposium from the Joseph P. Kennedy, Jr. Foundation, ed. G. A. JERVIS. Springfield, Ill.: Ch. C. Thomas 1968.
212. SIDMAN, R. L.: Development of interneuronal connections in brains of mutant mice, p. 163–193. In: Physiological and biochemical aspects of nervous integration, ed. I. D. CARLSON. Englewood Cliffs, N.J.: Prentice Hall 1968.
213. SIDMAN, R. L.: Cell interactions in developing mammalian nervous system, p. 1–13. In: Cell interactions, 3rd Lepetit Colloquium, ed. L. G. SILVESTRI 1971. 314 pp.
214. SPERRY, R. W.: Embryogenesis of behavioral nerve nets, p. 161–186. In: Organogenesis, ed. R. L. DeHAAN, H. URSPRUNG. New York: Holt, Rinehart and Winston 1965.
215. SPIEGEL, M.: The role of specific surface antigens in cell adhesion. I. The reaggregation of sponge cells. Biol. Bull. **107**, 130–143 (1954).
216. STEINBERG, M. S.: Does differential adhesion govern self-assembly processes in histogenesis? Equilibrium configurations and the emergence of a hierarchy among populations of embryonic cells. J. exp. Zool. **173**, 395–434 (1970).
217. STEINBERG, M. S., LEITH, A.: Biol. Bull. (in press) (1972).
218. STOKER, M. G. P.: Surface changes and growth of virus transformed cells, p. 271–282. In: Biomembranes, vol. 2, ed. L. A. MANSON. New York: Plenum Press 1971. 302 pp.
219. STOKER, M. G. P.: Tumour viruses and the sociology of fibroblasts. Proc. roy. Soc. **181**, 1–17 (1972).

220. Stoker, M. G. P., Rubin, H.: Density dependent inhibition of cell growth in culture. Nature (Lond.) **215**, 171–172 (1967).
221. Straznicky, K., Gaze, R. M.: The development of the tectum in *Xenopus laevis*: an autoradiographic study. J. Embryol. exp. Morph. **28**, 87–115 (1972).
222. Tal, C.: The nature of the cell membrane receptor for the agglutination factor present in the sera of tumor patients and pregnant women. Proc. nat. Acad. Sci. (Wash.) **54**, 1318–1321 (1965).
223. Tiedemann, H.: Factors determining embryonic differentiation. J. Cell Physiol. (Suppl. I) **72**, 129–144 (1968).
224. Townes, P. S., Holtfreter, J.: Directed movements and selective adhesion of embryonic amphibian cells. J. exp. Zool. **128**, 53–120 (1955).
225. Turner, R. S., Weinbaum, G., Burger, M. M.: (In preparation) 1972.
226. Uhlenbruck, G., Gielen, W., Pardee, G. J.: On the specificity of lectins with a broad agglutination spectrum. V. Further investigations on the tumor-characteristic agglutinin from wheat-germ lipase. Z. Krebsforsch. **74**, 171–178 (1970).
227. Vasiliev, J. M., Gelfand, I. M., Guelstein, V. I., Fetisova, E. K.: Stimulation of DNA synthesis in cultures of mouse embryo fibroblast-like cells. J. Cell Physiol. **75**, 305–314 (1970).
228. Vasiliev, J. M., Gelfand, I. M., Guelstein, V. I.: Inhibition of DNA synthesis in cell cultures by colcemid. Proc. nat. Acad. Sci. (Wash.) **68**, 977–979 (1971).
229. Vicker, M. G., Edwards, J. G.: The effect of neuraminidase on the aggregation of BHK21 cells and BHK21 cells transformed by polyoma virus. J. Cell Sci. **10**, 759–768 (1972).
230. Vlodavsky, I., Inbar, M., Sachs, L.: Temperature-sensitive agglutinability of human erythrocytes by lectins. Biochim. biophys. Acta (Amst.) **274**, 364–369 (1972).
231. Wallach, D. J. H.: Generalized membrane defects in cancer. New Engl. J. Med. **280**, 761–767 (1969).
232. Warren, L.: The biological significance of turnover of the surface membrane of animal cells, p. 197–222. In: Current topics in developmental biology, vol. 4, ed. A. A. Moscona, A. Monroy. New York: Academic Press 1969. 233 pp.
233. Warren, L., Critchley, D., Macpherson, I. A.: Surface glycoproteins and glycolipids of chicken embryo cells transformed by a temperature-sensitive mutant of Rous sarcoma virus. Nature (Lond.) **235**, 275–278 (1972).
234. Warren, L., Fuhrer, J. P., Buck, C. A.: Surface glycoproteins of normal and transformed cells: A difference determined by sialic acid and a growth-dependent sialyltransferase. Proc. nat. Acad. Sci. (Wash.) **69**, 1838–1842 (1972).
235. Warren, L., Glick, M. C.: Membranes of animal cells. II. The metabolism and turnover of the surface membrane. J. Cell Biol. **37**, 729–745 (1968).
236. Weinbaum, G., Burger, M. M.: Sponge aggregation. III. Isolation of a surface component required in addition to the aggregation factor. Biol. Bull. **141**, 406 (1971).
237. Weinbaum, G., Burger, M. M.: (In preparation) 1972.
238. Weiss, L.: The cell periphery, metastasis and other contact phenomena. Amsterdam: North-Holland Publ. 1967.
239. Weiss, P. A.: Cell contact. Int. Rev. Cytol. **7**, 391–423 (1958).
240. Weiss, P. A.: Neural development in biological perspective, p. 53–61. In: The neurosciences: Second study program, editor-in-chief F. O. Schmitt. New York: Rockefeller Univ. Press. 1970. 1068 pp.
241. Weiss, P. A.: Neuronal dynamics and axonal flow. V. The semisolid state of the moving axonal column. Proc. nat. Acad. Sci. (Wash.) **69**, 620–623 (1972).
242. Weiss, P. A.: Neuronal dynamics and axonal flow: axonal peristalsis. Proc. nat. Acad. Sci. (Wash.) **69**, 1309–1312 (1972).
243. Weston, J. A.: Proc. nat. Acad. Sci. (Wash.) (in press) (1972).
244. Wilson, H. V.: On some phenomena of coalescence and regeneration in sponges. J. exp. Zool. **5**, 245–258 (1907).
245. Wilson, H. V.: Development of sponges from dissociated tissue cells. Bull. Bur. Fisheries **30**, 1–30 (1910).
246. Wiseman, L. S., Steinberg, M. L., Phillips, H. M.: Experimental modulation of intercellular cohesiveness: Reversal of tissue assembly patterns. Develop. Biol. **28**, 498–517 (1972).

247. WOLPERT, L.: Positional information and spatial pattern of cellular differentiation. J. theor. Biol. **25**, 1 47 (1969).
248. WOLPERT, L., HICKLIN, J., HORNBRUCK, A.: Positional information and pattern regulation in regeneration of hydra. Symp. Soc. exp. Biol. **25**, 391–416 (1971).
249. WOLPERT, L., CLARKE, M. R. B., HORNBUSCH, A.: Positional signalling along hydra. Nature (Lond.) New Biol. **239**, 101–105 (1972).
250. WRAY, V. P., WALBORG, E. F., JR.: Isolation of tumor cell surface binding sites for concanavalin A and wheat germ agglutinin. Cancer Res. **31**, 2072–2079 (1971).
251. WU, H. C., MEEZAN, E., BLACK, P. H., ROBBINS, P. W.: Comparative studies on the carbohydrate-containing membrane components of normal and virus-transformed mouse fibroblasts. I. Glucosamine-labeling patterns in 3T3, spontaneously transformed 3T3, and SV40-transformed 3T3 cells. Biochemistry **8**, 2509–2517 (1969).
252. ZWILLING, E.: Limb morphogenesis. Develop. Biol. **28**, 12–17 (1972).

Namenverzeichnis / Author Index

Die *kursiven* Seitenzahlen beziehen sich auf das Literaturverzeichnis
Page numbers in *italics* refer to bibliography

Die in Klammern stehenden Ziffern beziehen sich auf die Nummern der Zitate innerhalb
des laufenden Textes und der Literatur

The numbers in parenthesis refer to the references in the text and in the bibliography

Sachverzeichnis / Subject Index

Ergebnisse der Physiologie, biologischen Chemie und experimentellen Pharmakologie

Reviews of Physiology, Biochemistry and Experimental Pharmacology

61. Band

1 portrait and 69 figures. III, 291 pages. 1969
Cloth DM 89,—; US $33.—

Contents: R. Jung: Paul Hoffmann, 1884 bis 1962. – W. Ulbricht: The Effect of Veratridine on Excitable Membranes of Nerve and Muscle. – M. Wiesendanger: The Pyramidal Tract. Recent Investigations on its Morphology and Function. – M. E. Holman: Electrophysiology of Vascular Smooth Muscle. – O.-J. Grüsser und U. Grüsser-Cornehls: Neurophysiologie des Bewegungssehens.

62. Band

1 portrait and 14 figures. VIII, 159 pages (23 pages in German). 1970
Cloth DM 48,—; US $17.80

Contents: M. Vogt: John Henry Gaddum, 1900–1965. – J. M. Marshall: Adrenergic Innervation of the Female Reproductive Tract: Anatomy, Physiology and Pharmacology. – W.-D. Thomitzek: Die Rolle des Carnitins im Intermediärstoffwechsel. – F. Bergel: Today's Carcinochemotherapy: Some of its Achievements, Failures and Prospects.

63. Band

79 figures. VIII, 241 pages. 1971
Cloth DM 98,—; US $36.30

Contents: H. O. Schild: Henry Hallett Dale, 1875–1968. – R. F. Schmidt: Presynaptic Inhibition in the Vertebrate Central Nervous System. – W. Schaper: The Physiology of the Collateral Circulation in the Normal and Hypoxic Myocardium. – H. Schmid-Schönbein, R. E. Wells Jr.: Rheological Properties of Human Erythrocytes and their Influence upon the "Anomalous" Viscosity of Blood.

64. Band

63 figures. VIII, 342 pages. 1972
Cloth DM 98,—; US $36.30

Contents: G. Moruzzi: The Sleep-Waking Cycle. – M. Jouvet: The Role of Monoamines and Acetylcholine-Containing Neurons in the Regulation of the Sleep-Waking Cycle.

65. Band

1 portrait and 45 figures.
VIII, 191 pages. 1972
Cloth DM 68,—; US $25.20

Contents: P. Kruhøffer, Chr. Crone: Einar Lundsgaard 1899–1968. – J.-S. Pitton: Mechanisms of Bacterial Resistance to Antibiotics. – J. B. Stanbury: Some Recent Developments in the Physiology of the Thyroid Gland. – M. A. Bouman, J. J. Koenderink: Psychophysical Basis of Coincidence Mechanisms in the Human Visual System.

66. Band

1 portrait and 46 figures.
VIII, 296 pages. 1972
Cloth DM 98,—; US $36.30

Contents: H.-J. Schümann: Peter Holtz, 1902–1970. – H. van den Bosch, L. M. G. van Golde, L. L. M. van Deenen: Dynamics of Phosphoglycerides. – N. Emmelin, U. Trendelenburg: Degeneration Activity after Parasympathetic or Sympathetic Denervation. – P. N. Patil, J. B. LaPidus: Stereoisomerism in Adrenergic Drugs.

67. Band

43 figures. IV, 226 pages. 1972
Cloth DM 86,—; US $31.90

Contents: K. Koizumi, C. M. Brooks: The Integration of Autonomic Reactions. – W. Elger: Physiology and Pharmacology of Female Reproduction under the Aspect of Fertility Control. – W. H. Daughaday, L. S. Jacobs: Human Prolactin.

Springer-Verlag
Berlin
Heidelberg
New York

München · London · Paris
Sydney · Tokyo · Wien

Ergebnisse der Physiologie

Biologischen Chemie und experimentellen Pharmakologie

Reviews of Physiology

Biochemistry and Experimental Pharmacology

Herausgeber / Editors

R. H. Adrian, Cambridge · E. Helmreich, Würzburg
H. Holzer, Freiburg · R. Jung, Freiburg · K. Kramer, München
O. Krayer, Boston · F. Lynen, München · P. A. Miescher, Genève
H. Rasmussen, Philadelphia · A. E. Renold, Genève
U. Trendelenburg, Würzburg · K. Ullrich, Frankfurt/M.
W. Vogt, Göttingen · A. Weber, Philadelphia

Reprint from Volume 68
P. A. Owren and H. Stormorken
The Mechanism of Blood Coagulation

Nicht im Handel

Springer-Verlag Berlin · Heidelberg · New York 1973

Inhalt/Contents

Ergebnisse der Physiologie

Biologischen Chemie und experimentellen Pharmakologie

Reviews of Physiology

Biochemistry and Experimental Pharmacology

Herausgeber / Editors

R. H. Adrian, Cambridge · E. Helmreich, Würzburg
H. Holzer, Freiburg · R. Jung, Freiburg · K. Kramer, München
O. Krayer, Boston · F. Lynen, München · P. A. Miescher, Genève
H. Rasmussen, Philadelphia · A. E. Renold, Genève
U. Trendelenburg, Würzburg · K. Ullrich, Frankfurt/M.
W. Vogt, Göttingen · A. Weber, Philadelphia

Reprint from Volume 68

R. S. Turner and M. M. Burger
The Cell Surface in Cell Interactions

Nicht im Handel

Springer-Verlag Berlin · Heidelberg · New York 1973

Inhalt/Contents

Ergebnisse der Physiologie

Biologischen Chemie und experimentellen Pharmakologie

Reviews of Physiology

Biochemistry and Experimental Pharmacology

Herausgeber / Editors

R. H. Adrian, Cambridge · E. Helmreich, Würzburg
H. Holzer, Freiburg · R. Jung, Freiburg · K. Kramer, München
O. Krayer, Boston · F. Lynen, München · P. A. Miescher, Genève
H. Rasmussen, Philadelphia · A. E. Renold, Genève
U. Trendelenburg, Würzburg · K. Ullrich, Frankfurt/M.
W. Vogt, Göttingen · A. Weber, Philadelphia

Reprint from Volume 68

W. W. Fleming, J. J. McPhillips, and D. P. Westfall
Postjunctional Supersensitivity and Subsensitivity of
Excitable Tissues to Drugs

Nicht im Handel

Springer-Verlag Berlin · Heidelberg · New York 1973

Inhalt/Contents

Handbook of Sensory Physiology

How is the information contained in an environmental stimulus converted into meaningful responses by an organism? How does an organism perceive the outside world, its energy fields in space and time, and transform it into purposeful reactions? These are central themes of sensory physiology. The Handbook of Sensory Physiology will compile authoritatively the present state of our knowledge in this regard.

Volume III, Part 1

Enteroceptors

Edited by Professor **E. Neil,**
Department of Physiology,
Middlesex Hospital Medical School,
London W. 1, England

With contributions by B. Andersson, M. Fillenz, R. F. Hellon, A. Howe, B. F. Leek, E. Neil, A. S. Paintal, J. G. Widdicombe

With 91 figures
VIII, 233 pages. 1972
Cloth DM 96,—
Subscription price
Cloth DM 76,80

The subscription price is applicable on orders for the complete set of published and unpublished volumes. All volumes and subvolumes are available separately at list price.

The book consists of six concise chapters on the functional characteristics of the various enteroceptors. Cardiovascular mechanoreceptors, arterial chemoreceptors, receptors of the lungs and airways and abdomino-pelvic visceral receptors represent the better known peripheral enteroceptors. Also included are two chapters on "central receptors" respectively concerned with thermoregulatory processes and with hunger and thirst.
Emphasis has been placed on the electrophysiological studies of these receptors and their afferent fibres but such emphasis has not excluded consideration of anatomical and electron microscope evidence of their site and nature. Similarly the role of these afferent endorgans in the body has been given due attention.
These essays are designed to interest the graduate student and to stimulate research. A full bibliography is provided after each chapter.

Contents

A. S. Paintal
Cardiovascular Receptors

A. Howe, E. Neil
Arterial Chemoreceptors

J. G. Widdicombe, M. Fillenz
Receptors of the Lungs and Airways

B. F. Leek
Abdominal Visceral Receptors

R. F. Hellon
Central Thermoreceptors and Thermoregulation

B. Andersson
Receptors Subserving Hunger and Thirst

Author Index
Subject Index

■ **Prospectus on request**

Outline

Springer-Verlag
Berlin
Heidelberg
New York

Universitätsdruckerei H. Stürtz AG, Würzburg